VENUS

About the Author

Anne Massey, C.A.P. (Certified Astrological Professional), is the International Vice President of the International Society for Astrological Research (ISAR) for Canada. She is also a member of the Finnish Astrological Association and the Canadian Association for Astrological Education. In 1991, Massey founded the Fraser Valley Astrological Guild, and serves as its current president. She has been a professional astrologer since 1988 and currently teaches, writes, lectures, and maintains a full-time practice.

A native of Finland, Massey has lived in Australia, the United States, and Canada, and enjoys networking with astrologers from all over the world. For the past eighteen years she has been researching planetary cycles, with special focus on Venus and Mercury. Her research and work on Venus combined with practical experience with clients form the basis of this book. She is currently working on her next book about Mercury cycles.

Her website, www.astrologicallyspeaking.com, features a collection of articles and essays based on her research, as well as her weekly horoscope column, which appears in print in the *Vancouver Courier* (British Columbia, Canada).

SPECIAL TOPICS IN ASTROLOGY

VENUS

♀

Her Cycles, Symbols & Myths

ANNE MASSEY

Llewellyn Publications
Woodbury, Minnesota

Venus: Her Cycles, Symbols & Myths © 2006 by Anne Massey. All rights reserved. No part of this book may be used or reproduced in any manner whatsoever, including Internet usage, without written permission from Llewellyn Publications except in the case of brief quotations embodied in critical articles and reviews.

First Edition
First Printing, 2006

Series design and format by Donna Burch
Cover art © Digital Archive Japan, Inc (DAJ)/Getty Images
Cover design by Ellen Dahl
Edited by Andrea Neff
Interior illustrations by the Llewellyn Art Department
Llewellyn is a registered trademark of Llewellyn Worldwide, Ltd.

Chart wheels were produced by the Kepler program by permission of Cosmic Patterns Software, Inc. (www.AstroSoftware.com).

Library of Congress Cataloging-in-Publication Data (Pending)
ISBN-13: 978-0-7387-0991-8
ISBN-10: 0-7387-0991-3

Llewellyn Worldwide does not participate in, endorse, or have any authority or responsibility concerning private business transactions between our authors and the public.

All mail addressed to the author is forwarded but the publisher cannot, unless specifically instructed by the author, give out an address or phone number.

Any Internet references contained in this work are current at publication time, but the publisher cannot guarantee that a specific location will continue to be maintained. Please refer to the publisher's website for links to authors' websites and other sources.

Llewellyn Publications
A Division of Llewellyn Worldwide, Ltd.
2143 Woodale Drive, Dept. 0-7387-0991-3
Woodbury, MN 55125-2989, U.S.A.
www.llewellyn.com

Printed in the United States of America

Other Books in Llewellyn's Special Topics in Astrology Series

Chiron, by Martin Lass
(2005)

Vocations, by Noel Tyl
(2005)

Eclipses, by Celeste Teal
(2006)

Houses, by Gwyneth Bryan
(2006)

Contents

List of Figures and Charts . . . xi
Foreword by Donna Van Toen . . . xiii
Introduction . . . 1

CHAPTER ONE—THE ESSENCE OF VENUS . . . 5
Keeping the Stories Alive . . . 6
The Goddess Venus . . . 7
Sleeping Beauty and the Little Mermaid . . . 8
The Other Goddesses . . . 10
Contemporary Venus . . . 15
Venus, Alive and Well . . . 16

CHAPTER TWO—THE BEAUTY OF VENUS . . . 19
The Basics . . . 20
Five Cycles . . . 21
The Precision of the Light of Venus . . . 22
Venus and Time . . . 23
The Rose and the Mathematical Formula for Beauty . . . 23

CHAPTER THREE—THE SACRED GEOMETRY OF VENUS . . . 27
Rose Pattern . . . 27
Divine Proportion . . . 28
The Golden Mean of the Motion of Venus . . . 29
Venus Cycle Dynamics . . . 30
The Synodic Cycle . . . 32
Quick Cycle Review . . . 33
The Pentagram: A Venus Cycle Symbol . . . 33
The Number Five . . . 35

The Venus Cycles . . . 37

Some Venus Stories . . . 41

Quick Review of Venus Cycles . . . 44

CHAPTER FOUR—VENUS STATIONS . . . 45

Retrograde Phenomenon . . . 46

In a Loop . . . 46

Up Close and Personal . . . 47

The Statistics . . . 48

Phase Change . . . 48

Retrograde Stations . . . 49

Hollywood Stories . . . 51

Real-Life Stories . . . 52

Direct Stations . . . 54

Fertility . . . 55

Marriage . . . 57

Individuals Born into the Venus Cycle . . . 57

Quick Review of Venus Stations . . . 59

CHAPTER FIVE—THE CYCLE OF HARMONY . . . 61

Awakening . . . 63

Inferior Conjunction . . . 64

New Venus . . . 65

Full Venus . . . 67

Venus Cycles in Your Chart . . . 69

Short Stories . . . 70

Your Venus Phase . . . 71

Another Technique to Determine Your Venus Phase . . . 72

My Thoughts . . . 72

CHAPTER SIX—WHERE DOES VENUS HOLD COURT? . . . 75

Angular Houses: Our Rituals . . . 77

Seventh House: Relationships . . . 81

Succedent Houses: Maintenance Power . . . 82
Second House: Value . . . 86
Cadent Houses: Knowledge . . . 88

CHAPTER SEVEN—VENUS IN ASPECT . . . 93
Venus Rulerships . . . 94
Easy and Challenging Connections . . . 96
Applying and Separating Aspects . . . 99
Declination . . . 99
Venus Out-of-Bounds . . . 100
Mutual Reception . . . 101
Venus Aspects . . . 101

CHAPTER EIGHT—CHARMING COSTUMES . . . 109
Venus in Earth: Taurus, Virgo, and Capricorn . . . 110
Venus in Air: Gemini, Libra, and Aquarius . . . 114
Venus in Fire: Aries, Leo, and Sagittarius . . . 117
Venus in Water: Cancer, Scorpio, and Pisces . . . 120
Quick Review of Venus in the Signs . . . 124
Two Wolves . . . 125

CHAPTER NINE—VENUS RETROGRADE: THE MAGNIFYING GLASS . . . 127
The Dance of the Seven Veils . . . 130
Likability Factor . . . 132
Venus Retrograde Statistics . . . 132
Closing Thoughts . . . 133

CHAPTER TEN—VENUS IN LOVE . . . 135
Venus in Aries . . . 137
Venus in Taurus . . . 138
Venus in Gemini . . . 139
Venus in Cancer . . . 140
Venus in Leo . . . 140
Venus in Virgo . . . 141

Venus in Libra . . . 142
Venus in Scorpio . . . 143
Venus in Sagittarius . . . 144
Venus in Capricorn . . . 145
Venus in Aquarius . . . 146
Venus in Pisces . . . 147
Closing Thoughts . . . 148

CHAPTER ELEVEN—VENUS AND MARRIAGE . . . 149
Marriage Indicators . . . 150
The Part of Marriage . . . 152
Married Venus . . . 153
What Might Venus Have in Her Arsenal? . . . 154
Bringing Out the Worst in Venus . . . 154
Closing Thoughts . . . 155

CHAPTER TWELVE—VENUS AND SOCIETAL CHANGE . . . 157
Gemini Occultations . . . 159
Sagittarius Occultations . . . 162

CHAPTER THIRTEEN—OTHER FEMININE ARCHETYPES . . . 165
Ceres . . . 166
Vesta . . . 168
Juno . . . 169
Pallas Athena . . . 170
Lilith . . . 172

Acknowledgments . . . 173
Appendix I: Astrological Data . . . 177
Appendix II: Venus Cycles Data . . . 181
Appendix III: New and Full Venus Listings . . . 187
Appendix IV: Four Pentagrams . . . 195
Bibliography . . . 199

Figures and Charts

1. Venus Symbol . . . 2
2. Five-Petalled Rose . . . 24
3. Pentagon and Decagon Superimposed on a Human Face . . . 25
4. Rose Pattern of the Venus Cycle . . . 28
5. Venus, Sun, and Earth Alignment at the Inferior Conjunction . . . 30
6. Venus Phases . . . 31
7. Synodic Cycle . . . 32
8. Pentagram with the Elements . . . 34
9. Venus Phase Cycle Pentagram . . . 36
10. Venus Synodic Cycle Pentagram . . . 38
11. Partial Chart for Client A . . . 42
12. Partial Chart for Prince Charles of England . . . 43
13. New Venus and the Retrograde Loop . . . 47
14. Retrograde Stations Pentagram . . . 50
15. Partial Chart for Peter . . . 52
16. Chart for Bill Gates . . . 53
17. Direct Stations Pentagram . . . 54
18. Chart for Julie . . . 62
19. New Venus Pentagram . . . 66
20. Full Venus Pentagram . . . 68
21. Angular Houses . . . 77
22. Succedent Houses . . . 83

23. Cadent Houses . . . 88
24. Venus/Mars Parallel . . . 94
25. Venus in Libra Aspects . . . 95
26. Venus in Taurus Aspects . . . 95
27. Rose . . . 136

Foreword

Drawing on the mythology of many cultures, the lives of some of our better-known cultural goddesses such as Marilyn Monroe, and the charts of everyday people both male and female, Anne Massey takes us on a journey of both discovery and rediscovery of the power of Venus in daily events and in the grand scheme of life in general. In doing so, she captures the voice of Venus as she expresses through the signs and elements, through the houses, and in aspect, giving Venus a life of her own. In fact, Venus was overseeing this birthing from start to finish. How do I know? Because when it came time for Venus in Pisces to speak, she was nowhere to be found in the manuscript. Yes, she had actually vanished! Now for those of you with Venus in Pisces, fear not. She was found and she will also speak to you. I mention this in passing mainly because it's such a wonderful example of what happens when you are truly in tune with the archetype you're writing about. And Anne is truly in tune.

You'll find all the cookbook basics here, and Anne has given them several new twists. You'll find contrasts between the various placements of Venus by element and by angular, succedent, and cadent house classifications. You'll find good aspect interpretations. There are chapters on Venus in love and Venus in marriage. There's even information about Venus in relationship to several major asteroids. You will discover that Venus can be naughty or nice, and you'll see where and how each facet of her multifaceted energy manifests.

But this book goes well beyond basics. You'll also learn about Venus as a timer and the eight-year cycle of Venus that brings us into harmony and awareness of our latent desires, helping us to externalize and manifest them. You'll learn to work with these cycles in your own chart too, using the well-laid-out tables in the appendix. And you'll learn what it means if your Venus is retrograde natally (and if you think it means you didn't play nicely with others in past lives, think again) or in a solar return.

Too often, Venus has gotten short shrift in the shuffle of astrological literature. We get lured away by the siren song of Neptune, mesmerized by the power of Pluto, intrigued by the surprising twists and turns of Uranus. How can a planet with the bland keyword of "nice" compete with these exotic energies? Anne will show you that it can and does. If love, money, social life, or enjoyment of the finer things in life interests you, you won't be taking Venus for granted anymore by the time you finish this book. Nor will you be bored as you read.

It is rare that I recommend a book as having something for everyone regardless of astrological level. This one does. Whether you're a new student or a seasoned pro, you can expect to come away with new information, along with an increased appreciation of and respect for Venus in all her guises and costumes—from sweetie pie to bitch goddess. This information will help you relate better to clients, friends, and yourself.

If you're ready to meet and make friends with Venus—the real Venus rather than the stick-figure one so often presented—read on. You won't be disappointed.

Donna Van Toen
Toronto
July 5, 2005

Introduction

Venus and Her Symbol

When we first study astrology, we learn that Venus represents love, relationships, and values. Each story or myth retells this, but naturally, with each new or revised version something changes . . . yet the essence remains. On the pages of this book, the spirit and heart of Venus are illustrated through myth, fairy tale, and contemporary Venus archetypes, while investigating this feminine archetype beyond Roman and Greek mythology.

The character and attributes of Venus help define how this archetype or actress plays out her role in our personal lives. Venus is also associated with beauty, fertility, protection, our ascension toward divinity, and feminine wisdom. She is the queen in our natal chart, and as we know from fairy tales, the queen character can have some nasty traits, including possessiveness, jealousy, and vindictiveness. Venus is also charming, beguiling, and enticing, and with her knack for diplomacy and cooperation can sweet-talk us into believing all kinds of things.

Figure 1—Venus Symbol

We will discuss Venus's placement and connections with stories of both famous and regular folk, and we will look at relationships and their timing. The symbol we draw to help us identify this archetypal energy is the same we draw in biology to represent a woman: the infinite circle resting on a cross. This symbol is known as the mirror. The circle has no beginning and no end; thus it represents eternity, infinity, and perfection—the spirit. The same shape is repeated in the disk of the Sun and used to represent God. The circle is a symbol of renewal and eternal return and is associated with the number ten.[1] The circular alchemical symbol of Ouroboros is a dragon or a snake biting its own tail. This symbol is used in a variety of cultures bearing different names. In African mythology, Ouroboros is symbolic of the creation myth. Astrologically speaking, we associate it with our cycles, the ecliptic, and various spheres within our solar system, as well as the horizon. In our standard wedding vows, we state, "With this ring, I thee wed"; the ring used is a symbol of unity and of something everlasting (". . . until death do us part"). A ring is also symbolic of possession and capture.

In the circular yin/yang symbol, we encounter the ebb and flow of day and night all within the great circle of the Tao. Tao means behavior, understanding, and the constant changing to and from. Taoism is a philosophy and a belief in simplicity and the very nature of the universe. There is no absolute stillness. Everything, including the universe, is changing all the time. Relative stability can be achieved when a state of harmony is reached between yin and yang, which are said to be the opposite but related natural forces in the universe.

The circle is also considered a symbol of protection and wholeness. The circle rests on the cross of matter, which holds the basic premise for the elements, and its symbolism represents our crucifixion into an incarnation. The elements of unity and separation are built into the cross, which is said to symbolize the cosmos. The horizontal line is the feminine principle, and the vertical line the masculine principle, and at their intersection point heaven and earth meet.

The angles of the cross are linked to the cardinal points of the compass and the four elements: fire, earth, air, and water. The horizontal line is also considered to symbolize time, and the vertical line eternity. Both parts of the Venus symbol are found in the astrological wheel divided by the Ascendant/Descendant line horizontally and the Midheaven (MC) and IC *(Imum Coeli)* axis vertically. No wonder we often refer to the Midheaven as the best we can become (or be), as it is the terminating or culminating point of the axis of eternity. The symbol for Venus, in its totality, represents bringing spirit into matter.

The element of duality is integral in astrology. When we first begin to learn astrology, we discover that half of the zodiac signs are feminine, and the other half masculine. In the astrological chart we have the masculine, outwardly manifested southern hemisphere at the top. The private feminine northern hemisphere, below the horizon of the circle, is separated by the line of time. The feminine western hemisphere and the masculine eastern hemisphere—separated by the line of eternity—repeat the same dualistic theme. Therefore, we need to remain cognizant of the philosophical axioms about polar opposites. Absorbing black (the absence of color) needs its complementary reflecting white (the presence of all colors); or in the most mundane of terms, the flower cannot bloom above ground without the root system buried within the earth. The most spiritual representation might be that we cannot be female incarnate without a male manifestation in the spiritual realm. Everything has a reverse or opposite side: love and hate, like and dislike, positive and negative. Our task is to learn to balance and integrate the various unique facets into our life and being.

Venus Cycles

All of the personal planets have potent cycles of long duration, which, astrologically speaking, are linked to significant times in our lives. The three personal planets—Mercury, Venus, and Mars—are intimately connected to our everyday lives and teach us about our ideas, values, and actions. Venus in our natal chart defines what we have learned thus far and what we are here to perfect in terms of our personal myths, beliefs, and values. Through the trials and tribulations of these Venus cycles, we further define our own estimation of ourselves within relationships and in our quest to support ourselves in the material realm.

The precision of the Venus cycle lends itself beautifully to helping us better understand ourselves and the process of change. Venus's cycles afford us incredible timing and predictive tools. We know that life is cyclical whether we utilize astrology or not. The cycles of the inner planets help us understand how we adjust and change over the course of time and in tune with the cycles.

The "dance of Venus" creates such ultimately positive and predictable changes in our lives that I know you will fall in love with Venus as I have over the years. All of the cycles of Venus will be illustrated in detail. Appendix I: Astrological Data provides a quick reference for Venus, and Appendix II: Venus Cycles Data includes data for her cycles.

---

1. In classical astrology, each of the seven planets known at the time has rulership over a certain number of years in the course of a lifetime, which was deemed to be seventy-five years. The Sun rules ten years of that time. The planetary order depends on whether the individual was born during the day or night. Venus rules for eight years, the Moon nine, Mercury thirteen, Mars seven, Saturn eleven, and Jupiter twelve. These periods are called firdaria and are similar to the Vedic dasha system.

CHAPTER ONE

The Essence of Venus

A word is not a crystal, transparent and unchanged; it is the skin of a living thought, and may vary greatly in color and content according to the circumstances and the time in which it is used.
—Oliver Wendell Holmes

We have been learning about Venus since childhood. The familiar words of the following nursery rhyme may not have been our first impression; however, this rhyme or a similar one was no doubt read to us at an early age.

> What are little girls made of?
> Sugar and spice,
> And everything nice,
> That's what little girls are made of.

Sugar and spice sounds better to a little girl than frogs and snails and puppy dog tails—the ingredients for little boys. The various Venus stories and mythologies illustrate similar principles. The nuances vary between cultures—time and language change our perceptions.

Thus, the newly revised contemporary versions reflect attitudinal changes in our society. This rhyme is such a memorable way to demonstrate the marked difference between the feminine and masculine. The Venus and Mars distinction takes place while we are young and impressionable. We learn early on that girls are different from boys.

Keeping the Stories Alive

Our lives—just like astrological lore—are filled with stories, myths, built-in dramas, and comedies as well as anecdotes.

Michael Grant and John Hazel define myths beautifully in the introduction to their book *Who's Who in Classical Mythology*: "Myths were invented to explain natural phenomena in a pre-scientific world, to elucidate sites and rituals and names of which the original meaning had been lost, to justify customs and institutions, to endow the gods with dramatic personalities and careers, to glorify nations and tribes and families and hierarchies and priesthoods, to fill out early history by inventive additions, to indulge wishful thinking by tales of adventure and heroism, and sometimes, merely to amuse and entertain: to beguile the long hours of darkness, or the tedium of a dusty journey, or a perilous tossing on the sea."[1]

In her paper *The Triangle of Love: Eros, Psyche and Poetry*, Carolyn Zonailo eloquently states: "Myths are metaphors—as such, they have been used throughout history to describe, analyze, categorize, and explain the essentials and vagaries of human nature. Myths attempt to go further than just the human—they exist to help us connect to the divine, the cosmos, and the unknown. Myths are the collective dreams of the ancestral soul."[2]

A myth is a traditional story concerning the early history of a people or explaining a natural or social phenomenon, typically involving the supernatural. Legends are traditional stories popularly regarded as historical, yet not authenticated. It is interesting to note that the origin of the word legend is from Latin—*legenda*, "things to be read." The word also has a second usage, allowing us to read a graph, image, or map, thus providing the key to opening up the secrets of the image, so to speak. The legends, stories, and myths hold keys to understanding Venus and her cycles.

Fairy tales are simplified, abridged introductions to complex myths. Rhymes, chants, songs, folklore, fairy tales, and poems represent the most traditional way of passing on knowledge and information. Short stories and rhymes are easy to memorize and to recall—

we can personally imagine and visualize the scenes and nuances. Some tales are universal, while others are culture-specific. Some of the best-known stories are the ones about Snow White, Cinderella, the Little Mermaid, and Sleeping Beauty. The original story lines have changed over the centuries, and some versions have been forgotten.

These stories originate in folklore and mythology, which are not copyrighted; thus, we are free to quote them as we remember them. This in itself promises that all the Venusian stories addressed in this book will automatically have personal perceptions and interpretations embedded. The fairy tales teach us about morals, magic, and human nature. The essence has endured, while the nuances have changed with each retelling of the story. Contemporary versions often alter the original dramatic ending, perhaps in order to make it more positive for the modern reader or viewer. Or could it be that the new version was adapted to reflect our changing societal views? The stories are alive—to live is to change.

The Goddess Venus

The leading lady of this book, Venus, whispers sweet nothings and mysteries of life with equal passion. Her name originates with the Roman goddess of love and beauty; however, the associated myth and lore are more ancient and found in most cultures. Before Greek mythology was incorporated into the Roman myth, the Roman Venus was merely a nymph in charge of gardens and fields. Nowadays when we read the story of Venus, it is identical to the one of the Greek goddess Aphrodite.

The Venus captured on canvas by the Italian painter Botticelli—the beauty standing on a half shell, the perfect woman with flowing hair covering the parts we are not comfortable viewing—was a latecomer. She displaced popular Roman goddesses such as Vesta, who is linked to the Greek Hestia, with a richer lore. The Romans had a goddess assigned to the Morning Star, Aurora, and Evening Star, Vesper, but these also became part of Venus. Venus quickly gained legendary proportions and even two different sets of parents. Rightly so, as the mythology was intended to define the energy that the planetary deity had, and with the elimination of earlier, more prominent goddesses, it was vital to assign these displaced qualities to an enduring archetype. Her glamour secured her elevation to the position she now holds; yet as an archetype, she is as old as time.

Not all paintings and sculptures inspired by Venus are demure. Venus herself was not depicted as demure in the mythological stories. Our view of Venus has become sanitized over the centuries, removing whatever we as a society are not comfortable accepting. This would naturally account for the fact that we still like to place women on pedestals where we can admire their beauty. Fashion magazines are a prime example. The faces of the women who grace the pages have been airbrushed to the extent that there appear to be no lines, creases, or imperfections, let alone a wrinkle or laugh line. With each brush stroke, a part of the woman's personality has been erased.

There are two stories of the birth of Aphrodite, and naturally, this beloved goddess has numerous names. In one version, she is the daughter of Zeus (Jupiter) and Dione, who was an earth goddess. In another version, she was born fully grown, emerging from sea foam as a result of Cronos (Saturn) having cut off the genitals of his father, Ouranos (Uranus), and tossed them into the sea. The name Aphrodite is typically translated as "born of sea foam." The latter is definitely not a version of her birth we would tell as a fairy tale to young ones. However, it is interesting to note that Venus was also considered the giver of life and the taker of life. The two names for these attributes of Venus, from Roman times, are Venus Libitina, associated with funerals and the extinction of life force, and Venus Genetrix, mother of the Roman people. Aphrodite emerges from sea foam, or spirit, into matter. We symbolically link the ocean to the unconscious or collective—the realm of Neptune. In fairy tales, the little mermaid becomes sea foam to be washed away by the ocean—ascension from matter into spirit.

Sleeping Beauty and the Little Mermaid

Before traveling back in time to meet the Babylonian Ishtar and the Sumerian Inanna, let us explore fairy tales. The story of the mermaid in *The 1001 Arabian Nights* has a devastating ending. In it, the daughter of Neptune, the king of the ocean, becomes sea foam after experiencing the love of a mortal man. The waves wash her away, never to be seen again. Contemporary versions are now told in movies such as Ron Howard's 1984 film *Splash*. The plot features elements of the original; however, the ending offers a different choice—the mortal man opts to join Madison, the mermaid, in the ocean.

We know that women are not perfect, and neither was Venus, nor the archetypes that preceded her and became a part of her story. The essence of the story of Sleeping Beauty is likely that of one of the Venusian goddess stories, only with imperfections and flaws removed. In order to understand perfection, we need to become intimately familiar with its opposite: imperfection. We are all flawed, otherwise why would we be here on Earth learning to improve ourselves? What we may not acknowledge is that we are perfect within. The polar opposites permeate philosophical and astrological texts. Ultimately, astrologers are very similar to philosophers, pursuing the intellectual logic and reason behind the planetary energies as these relate to the human experience.

In *Snow White*, we have all the elements of love, including jealousy, and the same rivalry between the Queen and Snow White as in the story of Aphrodite and Psyche. And in the end, both Snow White and Psyche get their prince. In the myth, Psyche regains her husband, Eros, who is Aphrodite's son, by proving through assigned tasks—often referred to as initiations—that she is worthy.

Let's explore what is in a name. The silver screen has immortalized Aurora as the best-known Sleeping Beauty. Aurora means the "goddess of the dawn," which is one of the many faces of Venus. In Finland, Sleeping Beauty is known as Princess Rose.[3] The Grimm Brothers wrote about the Briar Rose, the flower with five petals. The Rose is both symbolically and explicitly essential to Venus. This is illustrated in the chapter entitled "The Sacred Geometry." Imagine what it would be like to live surrounded by a beautiful rose fence that the world could not penetrate. In *Briar Rose*, Venus is also used as the name for one of the fairies. She ensures that the spell of eternal sleep can be broken. Some of the early versions were not quite as romantic, sweet, and innocent as the one by Walt Disney in 1959. The adultery, rape, and cannibalism included in the older versions have no place in a child's fairy tale.[4] However, these stories were not originally intended for children. Rather, they were a means of telling an embellished, real-life story in such a fashion that it could be thought to be the product of an active imagination. Beyond that, these fairy tales seem to retell ancient myths, as they were recalled and recorded at the time. What better way to keep the knowledge alive?

In each version, Sleeping Beauty is the most beautiful girl, who typically is pricked by a spindle and falls into eternal sleep. The spindle, like the turning wheel, marks the passage of time. Spinning, throughout history, has been at activity that women do—making clothes for the baby, the bride, and the shroud for burial. In *Sleeping Beauty*, the last godmother modifies the spell cast, reducing it to a length of one hundred years. The number of godmothers ranges from seven to thirteen; there are seven classical planets, ten in contemporary astrology, and thirteen constellations touching the ecliptic.

The lead character in each fairy tale naturally represents the feminine and specifically women. When we sleep, we are not conscious of our earthly realm. The figurative hundred-year sleep can be understood as a dream cycle or unconscious cycle during which we have no active control over our lives or destiny. The fact that it always lasts a hundred years in fairy tales is significant in terms of the Venus cycle, which ultimately is about awakening us to our spiritual nature. Each of the planets has a spiritual nature and purpose. In this book, the thread behind the discussion on Venus is about the awakening of the sacred or divine feminine through the symbolic essence of Venus.

We cannot write about Venus without using words and concepts that belong to her, such as girls, sugar, flowers, spices, rose, beauty, romance, and women. Even without any knowledge of astrology, we can see how these pretty words are Venus's domain. As astrologers, we often use shorthand, stating that a person is a certain way because she has Venus in Aries, for example. This wording invokes different imagery in my mind than in yours. I will discuss this particular placement of Venus later in this chapter. Words and what they mean change over time, and our understanding of their meanings also varies. Even beauty is said to be in the eye of the beholder; what is beautiful to one is not to another.

The Other Goddesses

The ancient goddesses listed here were more complex than and not as sanitized as the mythology of Venus. In order to accommodate the senses and sensibilities of people, many of our myths and fairy tales have undergone changes. Going back to the past, here are some of the names in mythology that are linked to Venus: Ana, Anat, Aphrodite, Astarte, Ashtar, Aurora (Morning Star), Shachar (Venus at dawn), Eos (Evening Star), Inanna, Ina, Ishtar, Freya, Frigga, Shalem (Evening Star), Sophia of Wisdom, Baalat-Gebal, and Hathor. Isis is also linked to Venus and to the Moon, and collectively called "she who was first." What is the essence of Venus? What are her many guises? That questioning awakened a persistent chant within me: *I am reborn in every woman.* She whispers to all of us, and poetry is within her realm.

> If someone painted a picture
> Of a Greek goddess in groves of green,
> It could never compare with you, my love.
> A beauty such as yours I have never seen.

> I know you're not the most beautiful in the world,
> But I don't only mean of face.
> There is much more in your mind, your expression, your charm, and your grace.
> You're beautiful; your face is so fresh,
> And each time I look at you it wouldn't be
> Too hard for people to guess
> That I am in love with you, no more, no less.

Rather than find a famous poem about beauty, grace, and charm, the attributes we assign to Venus, the above is an excerpt from a poem penned by my husband when I was sixteen.

The essence of Venus is in all of us, male or female; however, being in a female body is different from being in a male body. Here is an extract from Carolyn Zonailo's poem "The Tree of Life," published in her book *The Goddess in the Garden*:[5]

> In the beginning
> God created Adam
> From the mothering earth:
> His name means *man*.
>
> In the beginning
> God created Eve
> From the rib or side of Adam:
> Her name means *life*.
>
> Thus in the beginning
> God created male and female.
> And that is how we have existed
> Since the time of creation.

Eve is also a manifestation of the Venus archetype in one of her guises. According to biblical scholars, the first Eve, the one who refused to be in an unequal relationship with Adam, was Lilith. In the story of Inanna, she has Lilith, the dark maiden, banished from her Huluppu tree.

Research into the various Venus archetypes clearly shows that each culture had a goddess of love and beauty or a queen of heaven and night. Each archetype is linked to the fertility of both women and the land. In every myth, there were two other powerful women or personifications of a specific type of energy or characteristic associated with them. Typically, one was in charge of the underworld, which in psychological terms is associated with the unconscious. In myth, these strong characters represented disassociated facets of the Venus archetype. In contemporary astrology, the asteroid goddesses and Black Moon Lilith are becoming more important. We need new tools to discuss the changing role of women. Further, we are rediscovering the true essence of the feminine.

Who were the disassociated characters in mythology? Let us look at Psyche and Lilith. Aphrodite, the Greek personification of Venus, sent Psyche, her beautiful future daughter-in-law, on an errand to the underworld. Psyche needed to prove that she was worthy of Venus's son, Eros. In the Sumerian story of Inanna, Lilith was evicted from Inanna's Huluppu tree. Lilith is talked about as the first Eve, the one who considered herself equal to man. Lilith is typically associated with pure desire and sexuality, and to Inanna—the Sumerian Venus—this was a facet of Lilith with which she was not comfortable. Lilith emerges as the first completely independent female archetype. Chapter 13 addresses the four major asteroid goddesses and Lilith.

One role Venus plays is to awaken us to acknowledge our divine. In stories worldwide, divine wisdom is typically represented by a tree or something similar. The most poignant message seems to be the element of the quintessence; in Inanna's story, the word used for it was ME. This ME was invisible and never-changing and represented her invisible essence. The novel *The Initiation*, by Elisabeth Haich, is the story of a woman who recalls her past life in Egypt.[6] In it, she talks about ME in the same fashion, the part of her that will never cease to exist. The big Venus cycles, the hundred-year period and the predictable retrograde cycles, seem to serve the purpose of connecting us to our quintessence—our soul.

Those distant names echo in muted tones—Ashtar, Inanna, Frigga, Hathor . . . Who were they? Once upon a time, we worshipped these deities, built them temples, and offered tokens of our love and admiration. We carved their images on rock, sculpted their fertile forms, and captured their essence in copper and metal. In turn, our land and bodies were blessed with fertility. These immortal beings instilled in us an appreciation for beauty and security. Inanna was a powerful, temperamental goddess who grew into her role as the queen. She, like the other Venus archetypes, had a voracious appetite for men. The Sumerian poem is erotic and rather explicit. She took quick revenge against anyone who slighted her. A union between a king and one of her priestesses brought fertility to the land and power to the king.

Frigga, the beautiful Norse goddess, was wild and passionate—and associated with death in addition to love and beauty. Half of the souls of slain warriors belonged to her, and the other half to Odin, who, much like Mercury, can shift form and appearance. The Roman historian Tacitus associated Odin with Mercury. We know that Aphrodite had her affair with Hermes, as did Frigga with Odin. In Scandinavia, the constellation of Orion's Belt is named Frigga's Distaff—some believe the stars are her spinning wheel. Nowadays, spinning wheels are mostly antique decorations, but women still ensure that loved ones have the appropriate attire for each phase of life. Frigga's name also means beloved, and there are numerous spellings of her name. We have many terms of endearment for those we love.

The family tree of Inanna makes her the daughter of the Moon God and Moon Goddess, according to Wolkstein and Kramer.[6] In Egypt, Venus was known as Hathor, who was eventually eclipsed by Isis. To the Canaanites, the major goddess of fertility, sexual love, hunting, and war was the goddess Anat. In the Nordic countries, we encounter Freyja, and in the Celtic tradition, we have Brigid. Some of the deities and goddesses associated with the archetype of Venus are members of the original twelve gods and goddess of a pantheon. Originally, six were male and six female. However, we at times find that Venus is the one who claimed her rightful position among the major players; the best-known instance would be the Roman Venus.

The Babylonian creator goddess Ishtar—the giver of life—was associated with love, war, fertility, childbirth, healing, and justice. The main gate into Babylon was dedicated to Ishtar. In the book of Revelations 17:1–5, she is named the Great Whore.[8] The terms whore and harlot were used to describe groups with secret and occult knowledge. The word harlot nowadays refers to a particular kind of woman, but interestingly it used to refer to a particular kind of man—a man of no fixed occupation, a vagabond or beggar. For a time the word could also refer to a juggler or jester of either sex, but by the close of the seventeenth century, its usage referring to males had disappeared.[9] Words, concepts, and archetypes change over time, as does our understanding of these. We no longer refer to humans collectively as man or mankind. Political correctness has introduced a new word: humankind. The fascinating inference throughout the Venusian stories is that man sought divine knowledge from Ishtar and her counterparts.

Ishtar was an oracle and brought prophecy through dreams. Through her magic, others could obtain secret knowledge. She governed over sex and war, and protected men from evil. The Dance of the Seven Veils is a depiction of her descent into the underworld to bring back her husband. From Inanna, Ishtar, and Astarte through Venus, we have the symbolic description of her descent into the underworld and her journey back to the world. In all of these the myths, she is stripped of her seven veils and enters the underworld naked, vulnerable, and exposed.

The myths associated with Venus all describe her being stripped of worldly possessions and coming face to face with her other, hidden side. In the story of Inanna,[10] the Sumerian queen of heaven and earth, the other has a name, Ereshkigal, and is known as her powerful sister, who calls the underworld her domain. Inanna had seven disrobings at the seven gates. Seven is the number of planets in classical astrology. The crown was taken from her at the first gate. A crown is the symbol of sovereignty, distinction, and achievement. At the next gate her lapis beads were taken from around her throat. Lapis is considered to connect the heart and mind, ward off evil, and connect us with our higher self. The two strands of beads on her chest, the breast plate, the golden bracelet, her lapis measuring rod and line, and finally her robe—all of these were the royal symbols of her status. Inanna was judged, struck down, and hung on a hook like a piece of meat for three days and nights. The god of wisdom, Enki, fashioned two creatures to enter the underworld to rescue Inanna.

Raven Kaldera's book *MythAstrology* associates Venus in the signs with various feminine archetypes, providing some interesting points to ponder.[11] He associates Venus in Aries with Rhiannon, Taurus with Oshun, Gemini with Eros, Cancer with Mariamne, Leo with Ishtar, Virgo with Parvati, Libra with Aphrodite, Scorpio with Inanna, Sagittarius with Krishna, Capricorn with Freyja, Aquarius with Ganymede, and Pisces with Oba. Each of us recalls the stories differently; however, certain elements of the Venus myths resonate strongly with our understanding of a specific sign or house. We would all tend to agree that Venus in Scorpio would not be likely to seek superficial relationships. Assigning Aphrodite to Libra sounds perfect, as Libra loves harmony and beauty.

Contemporary Venus

If we were to define Venus through modern, contemporary women and their lives and images, whose image would we use? Marilyn Monroe died in 1961, yet her image has not. Posters bearing her image as a legendary beauty are still sold, and every year around her birthday, her life story is retold. Where does immortality begin? How can we consider her a mortal goddess without an exalted Venus? We feel we know a lot about her, yet very few of us actually knew her personally. Venus in her natal chart was in Aries in the ninth house trine Neptune, Ceres, and Lilith[12] in her first house. Venus is not in her natural element in Aries—the sign and house belong to Mars. Astrologically, we call this *detriment*, which means at a disadvantage, a loss, and originates with the Latin *deter*, which in turn means to discourage, prevent, and hinder action.

When Venus is a long way from home in the opposite signs of Scorpio and Aries, she is not as comfortable as she would be in Taurus and Libra, the two signs she calls home. Think about when we stay at someone else's home, and cannot behave as we can at our own place—we need to be on our best behavior and perhaps act differently than our norm. Another home Venus likes is Pisces, and likewise, the opposite sign of Virgo does not feel quite as comfortable to her. Venus in any sign is not good or bad. The sign placement of Venus simply helps define facets of our personality and attitude toward others. What we like and dislike are linked to this archetype in our chart.

The reason I began with the mortal goddess Marilyn Monroe is that she continues to be a Hollywood sex symbol despite the fact that she passed away decades ago. Perhaps we should pick Jayne Mansfield, Audrey Hepburn, Elizabeth Taylor, Jennifer Aniston, or Sarah Michelle Gellar instead. All of these women have Venus in Aries in their natal charts. Two of the women listed, Audrey Hepburn and Sarah Michelle Gellar, also have Venus retrograde in their natal chart. The placement of Venus in Aries certainly did not diminish these women's beauty, one of the more pronounced attributes we assign this complex inner planet. The women listed here may not have understood their natural allure. However, they were quite adept at becoming what was desired—accepting or inventing a glamorous facade. Venus is retrograde 7 percent of the time, a concept that will be discussed in chapter 9. Note that even though you may not have Venus retrograde in your natal chart, once every eighteen months you will get to experience these periods in your life. Could it be a natural step in our personal dance toward immortality—while we redefine our true values?

Why discuss Venus in Aries, a placement where Venus is living in a space that belongs to Mars? This sign is considered a weak or detrimental placement for Venus in classical astrology; however, her light is not diminished by this placement. Judging the strength of a planet using classical categorizations is valid in the right context, such as measuring the capacity to earn and hold onto money or the willingness to cooperate, yet Venus in Aries or in Scorpio—the natural domain of Mars—has the advantage of dwelling in Mars's territory. Here she is privileged to learn his ways and secrets and to know him intimately. Just think what you can learn about others if you are free to roam through their things. Our possessions do reveal a lot about us. Women with Venus in Aries or Scorpio know men well and can naturally push their buttons with flair. Unlike the many naked-yet-covered portraits of Venus, Venus is not all demure—especially Venus in these two signs. A similar theme applies to men who have their Mars in a Venus-ruled sign (Taurus and Libra) and men with Venus exalted in Pisces. These men seem to be truly in touch with the feminine.

Venus, Alive and Well

Words and concepts are living entities, which in astrology are represented by planets, signs, houses, and various other points. Astrology reflects life on earth and our personal experience of it. Planets impel rather than compel, and in time, we heed their counsel. Our astrological chart is alive in the same manner that we are alive.

Venus is alive and well, and her story forms an intriguing thread in contemporary literature. Both historical and fictional books centered on Venus have become increasingly popular in the twenty-first century. The hottest seller in recent times was Dan Brown's *The Da Vinci Code*, in which he talks about Venus revealing some so-called secrets—Venus brought to the masses.[13] Christopher Knight and Robert Lomas have published several books on the Freemasons, which discuss the Venus cycle extensively.[14] What is truly fascinating about these timely books is that Venus is taking center stage through mainstream publications. Astrological texts have not held this kind of position, nor do they typically hit the bestseller lists. Venus, as a result, is now gaining a bigger reputation once again with strong linkage to biblical feminine figures. As a society, we prefer reading a story, yet this beguiling archetype keeps finding new ways to stay in our consciousness. My perception is that she is all around us—in novels, movies, nursery rhymes, and the increasingly popular contemporary retellings of legends.

As a final thought, the following is an excerpt from Dan Brown's novel: "Langdon quickly explained that the Rose's overtone of secrecy was not the only reason the Priory used it as a symbol for the Grail. Rosa Rugosa, one of the oldest species of rose, had five petals and pentagonal symmetry, just like the guiding star of Venus, giving the Rose strong iconographic ties to womanhood. In addition, the Rose had close ties to the concept of 'true direction' and navigating one's way. The Compass Rose helped travelers navigate, as did Rose Line, the longitudinal lines on maps. For this reason, the Rose was a symbol that spoke of the Grail on many levels—secrecy, womanhood and guidance—the feminine chalice and guiding star that led to secret truth."[15]

---

1. Michael Grant and John Hazel, *Who's Who in Classical Mythology* (1973; reprint, New York: Routledge, 2003), p. vii.
2. Lecture notes. Carolyn Zonailo can be contacted at http://www.carolynzonailo.com.
3. Sleeping Beauty's earliest modern influence appears to come from *Perceforest*, an Arthurian romance that was first printed in 1528. Heidi Anne Heiner, "History of Sleeping Beauty," *SurLaLune Fairy Tales*, http://www.surlalunefairytales.com/sleepingbeauty/history.html.
4. Ibid.
5. Carolyn Zonailo, *The Goddess in the Garden* (Victoria, B.C.: Ekstasis Editions, 2002).
6. Elisabeth Haich, *The Initiation* (Santa Fe, NM: Aurora Press, 2000).
7. Diane Wolkstein and Samuel Noah Kramer, *Inanna, Queen of Heaven and Earth: Her Stories and Hymns from Sumer* (New York: Harper & Row, 1983).
8. Revelations 13:4—modern versions of the Bible only talk of the City of Babylon.
9. *American Heritage Dictionary of the English Language*, fourth edition. Copyright © 2000 by Houghton Mifflin Company.
10. For the hymns and commentary of the original text, please refer to *Inanna, Queen of Heaven and Earth: Her Stories and Hymns from Sumer,* by Wolkstein and Kramer (New York: Harper & Row, 1983).
11. Raven Kaldera, *MythAstrology* (St. Paul, MN: Llewellyn Publications, 2004).
12. I will simply use Lilith when talking about Black Moon Lilith. There are two other Liliths in our astrological symbolism, which I will not discuss in this book.

13. For example, see page 202 of *The Da Vinci Code*, by Dan Brown (New York: Doubleday, 2003).
14. Christopher Knight and Robert Lomas, *Uriel's Machine: The Ancient Origins of Science* (London: Arrow Books, 2000) and *The Hiram Key* (London: Arrow Books, 1996).
15. See reference 13.

CHAPTER TWO

The Beauty of Venus

In mid-December 1989, five female clients with marriage issues booked individual appointments with me in the course of two days. Each one wanted to discuss divorce, feeling that love had left their marriage. Venus was in Aquarius at the time. Uranus, the modern ruler of Aquarius, rules divorce. The appointments were made for January 1990. Back then, all of this astrologer's delineations were painstakingly typed—preparation took a long time. The fact that Venus was stationing retrograde prior to the New Year prompted a call to an older, more experienced astrologer, who specializes in horary astrology.[1] What to make of it? She said that because Venus was retrograde for all of January, one thing that was quite certain was that there would be no divorce. A fixed Venus holds on, and even tighter when she is retrograde.

She was right—none of the women actually divorced. One cancelled her appointment, and the other four reported back six months later with an update—feedback is a wonderful learning tool. Each woman found a way to discuss the issues honestly and openly with her husband. The heart-to-heart conversations resulted in a new and better arrangement. Venus retrograde periods offer us an opportunity to assess what we need to do, so that we love what we have or are doing.

This experience launched a sixteen-year period of research into Venus cycles. Initially, the retrograde stations were the primary focus; after all, these leave a strong imprint on our charts. On a personal note, the Venus retrograde station in 1989 was square my natal ninth-house Sun,[2] opposite my eighth-house Uranus, and conjunct Chiron in my second house (birth at high latitude). Nine months later, we sold our house and bought a new one, as my husband was transferred to the Canadian office of his company. I was busy starting an astrological association and establishing my practice. Four years later, with the Full Venus in Capricorn, which repeated the New Venus degree of 1990, I stepped back from active organization of the association's events and began a second career in accounting (second house). The 1998 New Venus in Capricorn brought about a return to running the astrological association once again. I had spent the four months prior to that feeling unsatisfied with my studies for certification as an accountant, and I no longer enjoyed my other job as a baseball umpire. I also found that I missed my involvement with other astrologers. I dropped out of the course, gave up umpiring, took up astrology with full concentration once again, and knew I was doing what I truly loved. The beauty of these eight-year cycles is that they gently bring us closer to our heart's desire.

The Basics

In regular motion, Venus breezes through a sign in about twenty-five days. She forms aspects to our natal planets for two or three days at a time, resulting in a few compliments, a couple of kudos, and perhaps some nice times with others. After all, she is one of the two benefic planets in astrology. Jupiter is known as the greater benefic and Venus as the lesser one. In the context of transits, Venus stays with a personal planet for a shorter period of time than Jupiter, thus "lesser" seems apropos. The word benefic has some interesting concepts built into it. The word benefic comes from the Latin *beneficus*, which means to do something well, do a good deed; helping others, especially by donating for charitable purposes; the money, or help given. Our culture has translated this benefic influence to acquiring rather than giving.

Venus is never farther than 48° from the Sun; therefore, our natal Venus can never be too far from the planet that represents our vitality, ego, and will. Venus always retains the rays and heat of the Sun. Let's look at an example to illustrate where Venus could be in relation to the Sun in a specific degree. If the Sun is at 15° Cancer, Venus can be positioned at 28° Taurus or at 2° Virgo. Naturally, we can find Venus at any position between 28° Taurus and 2° Virgo when the Sun is at 15° Cancer. When a planet is close to the Sun, it loses part of its identity to the Sun. In the case of Venus, when she is in the presence of the Sun, our desires and social skills need to be blended with the individual and ego demands of the Sun.

The only major aspect Venus can form with the Sun is the conjunction. The exact aspect occurs once every nine months; however, these conjunctions take place only in one of five zodiac signs in any given year. At the beginning of the twenty-first century, these are Aries, Gemini, Leo, Scorpio, and Capricorn. At the beginning of the twentieth century, they were Taurus, Cancer, Virgo, Sagittarius, and Aquarius. We can see that there is a hundred-year theme. In addition, we have a repeat of each hundred-year cycle through a zodiac sign after a quarter of a millennium has passed. For example, the first New Venus in Scorpio took place on November 22, 1930, at 29°44' Scorpio. The last one out of the thirteen will take place on October 24, 2026, at 0°45' Scorpio. The next New Venus in Scorpio will take place on November 22, 2181, at 29° Scorpio. Venus works her magic through degrees.

Five Cycles

Venus is not merely a personal planet that marks time in days or weeks. If we look up the Sun-Venus conjunctions in one of the five signs—Aries, Gemini, Leo, Scorpio, and Capricorn—we find one every four years. Every second time Venus is retrograde at the conjunction. These are the inferior conjunctions, or the New Venus. "Inferior" refers to the fact that the path of Venus lies between the Sun and the Earth—Mercury and Venus are the two inferior planets. The New Venus in the same sign repeats every eight years. The Full Venus in the zodiac sign also repeats every eight years. If we count the number of conjunctions in an eight-year period, we find thirteen when Venus is direct and thirteen when she is retrograde. Eight years times thirteen conjunctions makes 104 years. The conjunctions when Venus is retrograde are referred to as New Venus in this text. The conjunctions of the Sun and direct Venus are referred to as Full Venus. The New Venus emerges born within the depths of her cycle marking a new beginning. The term inferior conjunction fails to illuminate that a waxing phase has begun. The term Full Venus illustrates that Venus is glowing under the Sun's rays.

If we thumb through the ephemeris, the patterns do not emerge with ease. However, check out your own birthdays for the past forty or fifty years. Theoretically, we should find Venus in the same sign as at birth every eight years. If you were born with Venus in a sign other than Aries, Gemini, Leo, Scorpio, or Capricorn, you will find that Venus is in your natal sign as frequently as every third or fourth year.[3] If you were born with Venus in one of the five zodiac signs, Venus is in the same sign as at birth on your birthday every eighth year retrograde in motion, and the following year she is direct in the natal sign. Every eighth year Venus will be conjunct the Sun on your birthday.

Solar returns are charts we cast for a person's birthday. In a solar return chart, the Sun is always at the same degree as in our birth chart. We might expect Venus to be retrograde in the chart once every eight years. This only applies to those of us born with Venus in one of the five signs where she retrogrades during our lifetime. Those who have Venus in a sign where she does not retrograde may have to wait until they are thirty years of age before experiencing a retrograde Venus in the solar return chart.

Checking out Venus's position on your birthdays may reveal a set of repeating degrees—every eight years. Venus naturally emphasizes degrees, and this theme can add an intriguing dimension to the natal chart. This technique is perhaps a little too labor-intensive to use with clients, but with our own charts this offers additional insight into why certain degrees seem so sensitive in our charts. For example, you may notice that the degree of Venus in the solar return chart on your birthday when she is retrograde is within a degree or two of the one for the following year. During the years Venus is retrograde in the solar return, we may find that money matters are tough. The following year, Venus will be direct in the same spot, and those obstacles are removed.

Each of the five cycles encompasses a retrograde station and a direct station, which repeat the degrees. We may not respond to all of them, but eventually our natal chart is triggered by at least one of the Venus cycles. The waxing New Venus becomes a Full Venus after nine months, and each New Venus culminates to full at the about the same degree four years later. Eight years after that, the New Venus begins at a point about 2° earlier in the same sign. Venus measures time with the precision of a metronome.

The Precision of the Light of Venus

There is an additional eight-year cycle of Venus. This research comes from Christopher Knight and Robert Lomas, the authors of *Uriel's Machine*.[4] Part of their research is on the megalithic structures in Ireland. Just north of Dublin lies Newgrange. Scientists have long known that one of the openings in the structure at Newgrange allows the rays of the setting winter solstice Sun to enter the inner chamber. There is a second opening, which is more intriguing. Every eight years, twenty-four minutes after the sunset, the reflected light of Venus enters the inner chamber of this megalithic structure, which has been dated to approximately 5,000 years ago. At this point the Sun is at 23° North 27' and Venus at 23° North 18', situated at approximately 25° Sagittarius—the Galactic Center. The Sun is ahead of Venus in longitude, thus she is a Morning Star. The years given by Lomas and Knight are 1993, 2001, 2009, etc. Remember to add and subtract in increments of eight; however, the date is always the winter solstice. This is the fact from which the inference to the metronome of our solar system is drawn.[5] Venus repeats each action in the same shape and form, or position and location, at eight-year intervals without fail.

Children conceived at the spring equinox would have been born at the winter solstice. A child born at the time the Venus light entered this chamber at Newgrange was thought to have a direct channel for reincarnation. Perhaps it was believed that leaders were born with the light of Venus emanating from the Galactic Center.

Venus and Time

The megalithic structures in Ireland are older than the pyramids in Egypt. History is written in hindsight and represents the consensus at the time it was recorded. New discoveries are being made at incredible speed, due in part to the advancement of our technologies. However, the Dresden Codex, which illustrates the Mayan calendar and was probably written just before the Spanish conquest, is an incredibly accurate calendar that has dazzled scholars with its precision. The cycles of Venus observed by the Mayans are central for timekeeping. According to Knight and Lomas, "The movements of Venus—the metronome of the Zodiac—are far more reliable than the proverbial Swiss clock. If one understands the position of Venus, one knows the time and date to a precision measured in seconds over hundreds of years."[6] Prior to the development of atomic clocks in the twentieth century, astronomers used the Venus cycle to correct the civil calendar.

The Rose and the Mathematical Formula for Beauty

Venus is linked to time, harmony, and beauty. We also associate flowers with Venus, in particular the rose, which we offer as a token of our love. In nature, flowers with five petals, such

Figure 2—Five-Petalled Rose

as the wild rose, are always the uncultivated version. The rose is a symbol of many sacred and secret orders and has often been used in reference to women in general. This beautiful, fragrant flower is also a symbol of rebirth and resurrection when it is red. A white rose is considered a flower of death and sorrow. The pink one is associated with budding love, and in Victorian times, the message of the yellow rose was "forget me." The messages behind the choice of flowers have been lost in our times. The Mystic Rose, according to Lomas and Knight, is a reference to the mystical properties of Venus, which were central in understanding the occult or hidden mysteries of nature and science, a fundamental theme in Freemasonry.[7]

Venus still delivers messages in the shape of the rose. The pattern of her cycle prints a five-petalled rose in the heavens; the completed rose emerges once every eight years. The geometric ratios of her dance relative to the Sun and the Earth are in a beautiful harmony. Those ratios are present in our faces and bodies and in nature.

Venus holds the formula for beauty—the Greeks were credited with knowing the mathematics behind it. For example, we consider teeth beautiful when the length of the two front teeth divided by their width equals the golden mean. We have numerous other words for this ratio: golden or divine rectangle, ratio, and proportion. The forearm and hand ratio and various proportions in our bodies also conform to this divine proportion. In order for our upper body to be in harmonious proportion, the width of our shoulders should be 1.6 times the width of our waist. The measurement from the inside corner of an eye to the tip of our nose, and the measurement from the inside corner of an eye to the corner of our mouth, also need to be in this ratio. (The first measurement, when multiplied by 1.6, should give us the second measurement.) The head forms a golden rectangle with the eyes at the midpoint. The width of the face is 1, and the height is 1.6. By drawing a line to divide the rectangle in half horizontally, we arrive at that midpoint.

Figure 3—Pentagon and Decagon Superimposed on a Human Face

The shape of the pentagon with five angles, when superimposed on a human face, defines what we see as beautiful. If the golden proportion, present in the pentagram shape with its five angles, is present in the features of the face and body, we see these as beautiful. This geometric shape is used to define the beauty in proportion. We may think this is ancient history and no one uses it any longer. Apparently, not so! There are cosmetic surgeons who have created a mask using two geometric shapes—a pentagon and decagon—as a template to determine which features need to be improved to create a beautiful face. A decagon is made from two pentagons. The Venus cycle symbol discussed in the next chapter features this shape.

The heart shape used to express love bears no resemblance to the physical heart, a fist-shaped muscle ruled by Leo. Did you know that two perfect circles and one square melded together are the building blocks of the heart of love? These circles, with their inherent symbolism, can be seen as a representation of the two individuals bonded together on Earth by time and eternity. The beauty and innate symbolism in geometry is fascinating. This geometry is present in the rhythm of the Venus cycle, so let us begin to explore it in intimate detail.

---

1. Horary astrology answers questions based on the chart cast for the time at which the question is posed.
2. Birth data: October 29, 1955, 12:50 p.m., Helsinki, Finland (NKL), from hospital records.
3. • The retrograde station signs are Aries, Gemini, Virgo (moving into Leo in 2022), Scorpio, and Aquarius (moving into Capricorn in 2013).
   • The New Venus signs are Aries, Gemini, Leo, Scorpio, and Capricorn.
   • The direct station signs are Aries, Gemini, Leo, Libra (the last direct station in Scorpio was in 2002), and Capricorn.
   • The Full Venus signs are Aries, Gemini, Leo, Scorpio, and Capricorn.

   By the time you have finished reading the cycle chapters, the idea of the motion backward through a century, the precise ebb and flow of Venus, will be etched in your mind. If a station or conjunction is early in the zodiac sign, the sign changes in eight, sixteen, or twenty-four years from now. If the station or conjunction is in a late degree, it will be close to a hundred years before we see a change. Venus retrogrades against the order of the zodiac. The New Venus in early Scorpio cycle will become the New Venus in late degrees of Libra by 2032. The cycle in each of the five signs moves back about 2.5° at eight-year intervals. If you look at the four listings in this endnote, you will see that the final retrograde station in Aquarius took place in 2005—the very theme that launched my personal research into the Venus cycles.
4. Christopher Knight and Robert Lomas have published several books on their research into the origins of science, including *Uriel's Machine*, *The Hiram Key*, *The Second Messiah*, and *Unlocking the Hiram Key*. See http://www.knight-lomas.com.
5. Metronomes were designed to help the sense of timing and tempo when playing music and to help develop a sense of rhythm. The rhythm of our heart is between 60 to 100 beats per minute when we are resting and varies by age and fitness level. Ballads and the waltz are played at the tempo of about 40 to 60 beats per minute, Motown is 80, and rock and roll starts at 100 and goes up to about 210.
6. Christopher Knight and Robert Lomas, *Uriel's Machine: The Ancient Origins of Science* (London: Arrow Books, 2000), p. 108.
7. Knight and Lomas, *Uriel's Machine*.

CHAPTER THREE

The Sacred Geometry of Venus

> *. . . there is music wherever there is harmony, order and proportion;*
> *and thus far we may maintain the music of the spheres; for those well ordered*
> *motions and regular paces, though they give no sound unto the ear,*
> *yet to the understanding they strike a note most full of harmony.*
> —T. E. Browne

The sacred geometry of proportions, visible in patterns and designs of organic life, is the natural rhythm of planetary cycles. The dance of Venus with the Sun and Earth has a tempo not unlike the waltz—eight to thirteen, which is known as the golden or divine proportion; that is, eight years and thirteen stations or conjunctions. The waltz is written to 3/4 time or 9/12; the Venus rhythm is just slightly off that at 8/13. The innate rhythm of the cycles sets the natural beat of life; there is a perfect time for everything. The incredible symmetry, the simplicity, and the precise, predictable timing of the Venus cycles are truly a work of art.

Rose Pattern

A picture is worth a thousand words. Over a period of eight years, Venus traces the rose pattern in her dance with the Sun. Figure 4 is an illustration of the astronomical plate

Figure 4—Rose Pattern of the Venus Cycle

from James Ferguson's *Astronomy Explained upon Sir Isaac Newton's Principles*, 1799 edition, Plate III. The little circle in the center represents the Earth; the patterns represent the motion of Venus.

The symmetry of Venus's cycles within cycles is breathtakingly beautiful. The five consecutive synodic cycles of Venus complete the rose shape. The word synod means "to come together." The loops are formed by her retrograde periods, which take place at approximately 584-day intervals. The tip of the petals represents the Full Venus, and 584 days later she is painting another tip.

Divine Proportion

The golden mean or rectangle is encountered when taking the ratios of geometric figures, such as the pentagon (five angles), decagon (10 angles) and dodecagon (12 angles). Plato and Pythagoras, in their mystery schools, considered the divine proportion, or golden mean (1.618) sacred. There are five cycles of Venus every eight years. Eight divided by five equals 1.625. The numerical value of the golden ratio is unending and usually rounded off to three decimal places. Italian mathematician Fibonacci, who lived around 1175–1250

CE, gave us the Latin numbers, the Fibonacci series, and the decimal system as opposed to the Roman numerals in use until the thirteenth century. The Fibonacci series of numbers is 1, 2, 3, 5, 8, 13, 21, 34, 55, 89, 144 . . . The addition of any two consecutive numbers results in the next number in the sequence. The ratio of two consecutive numbers in the series is the golden mean. As the numbers in the Fibonacci series increase, the decimal representation grows higher.

Most life forms, including DNA strands, reflect this divine proportion. In medieval times, buildings were created using this harmonic ratio. Some shapes look more pleasing—it is all in the proportions. No wonder mathematicians are so excited about their work. In astrology, we assign harmony and beauty to Venus, but we also need to add the attributes precision, proportion, and timing.

In medieval times, a dressmaker had to be familiar with proportions in order to respect the standards of modesty. The dressmaker was allowed to measure the woman's wrist and then needed to know how to calculate her other dimensions from that information. Somewhere along the way, we lost touch with those skills that used to be commonplace. Imagine walking into a fashion house today, offering the circumference of your wrist as a measurement, and ordering a gown!

The Golden Mean of the Motion of Venus

That geometric ratio present in a beautiful body and face is also present in the motion of Venus. Venus orbits the Sun in 224.695 days, and the Earth orbits the Sun in 365.242 days, creating a ratio of 8/13. The ratio of the orbit of Venus in her dance with the Sun and the Earth equals the golden mean of 1.625—13 divided by 8. Five conjunctions of Earth and Venus take place every eight orbits of the Earth around the Sun and every thirteen orbits of Venus. In classical astrology, conjunctions were excluded from the aspect family, as they represent a union, or a syzygy (which also means a union). The original meaning of the word aspect meant to look at or to see—planets next to each other are not gazing at one another.

Astronomically speaking, the term conjunction can also be used to describe an opposition of two astronomical bodies. We measure new cycles between the Sun and planets from the point where the planet is closest to Earth. Figure 5 shows the inferior conjunction, or New Venus—the three planets are aligned. Venus lies between the Sun and the Earth.

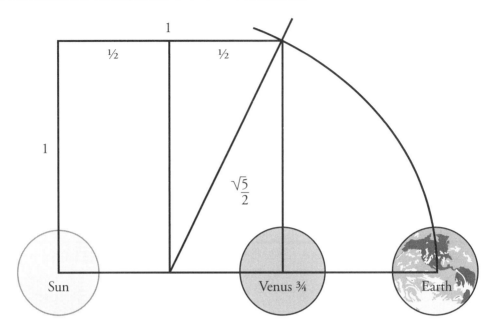

Figure 5—Venus, Sun, and Earth Alignment at the Inferior Conjunction

Mercury and Venus are the only two inferior planets, because their paths around the Sun lie between that of the Sun and Earth. At the Full Venus, the Sun and Venus trade places—Venus is on the far side of the Sun and farthest from the Earth. Let us look at a few more illustrations to see how the phases of Venus operate.

Venus Cycle Dynamics

As Venus travels around the Sun, different illuminated areas are visible from the Earth, making it look as if Venus is changing shape and size. From this perspective, Venus can be seen on the opposite side of the Sun about every 584 days. This is the superior conjunction, which takes place when the Sun is between the Earth and the inferior planet.

Approximately 221 days after the Full Venus, or the superior conjunction, half of Venus is illuminated. Add another 71 days, making it 292 days total, and only a crescent is illuminated by the Sun. Venus is on the same side of the Sun as the Earth at the 292-day mark. Each phase—Morning and Evening Star—is 292 days long on average. Venus does not move at uniform speed. On either side of the Full Venus, Venus speeds along at 1°15'

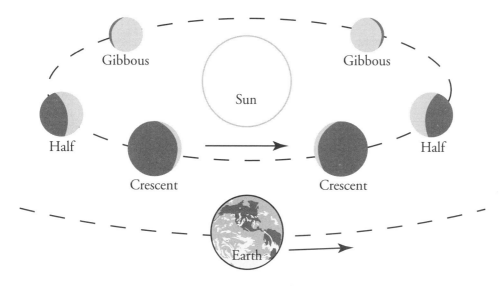

Figure 6—Venus Phases

per day and then slows down about 90 days later. For several months prior to the retrograde station, she moves at a continuously decreasing pace. Note that the quarter phase takes place 146 days after the Full and New Venus.

Figure 6 shows the relationship of the Sun in the middle with Earth at the bottom. Between the two gibbous phases, Venus is at the top (not shown), fully illuminated by the Sun. At this point, the planet Venus is at her apogee, i.e., farthest from the Earth. The superior conjunction is similar to the Full Moon in astrological significance, thus I refer to it as the culmination. Astrologically speaking, at both the New and Full Venus—or the inferior and superior conjunctions—the Sun and Venus are together in the sky from our Earthbound perspective, not opposite one another. Our astrology is Earth-centered, and we look at everything from our vantage point.

The duration of each phase—waxing and waning—is approximately as long as a human pregnancy, i.e., 280 days, forty weeks, or nine months. During the transition from Evening Star to Morning Star, Venus spends forty days and forty nights moving retrograde from our Earthbound perspective and is closest to the Earth.

Venus is visible up to three hours around sunrise as a Morning Star, and for the same length of time at sunset as an Evening Star. When Venus is higher in longitude than the

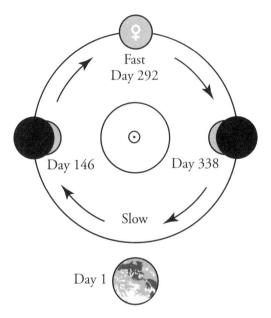

Figure 7—Synodic Cycle

Sun, in the order of the zodiac, she is an Evening Star. As an example, Venus at 25° Aquarius is an Evening Star when the Sun is at 5° Aquarius. Should the Sun be at 8° Pisces instead, Venus would be a Morning Star rising before the Sun. To see this visually in a chart, turn the wheel so that the Sun is at the nine o' clock position. Venus below the horizon is an Evening Star, and above the horizon—rising before the Sun—a Morning Star.

The Synodic Cycle

The inferior conjunction launches a new synodic cycle between Venus and the Sun. Astronomically, the word synodic relates to the alignment of celestial bodies or the interval between occasions when the same celestial bodies are aligned. Religious organizations use the word to describe their recurring meetings.

Figure 7 illustrates the waxing and waning cycle of Venus. Day 1, at the bottom of the graph, is preceded by Day 584. Venus moves slowly when she is near the Earth and fast around the Full Venus. The New Venus period is also referred to as the heliacal rising—she rises with the Sun. The Evening Star Venus is called the heliacal setting. Helios is the Greek word for the Sun.

The Evening Star appears to vanish during the retrograde period before emerging as the brilliant Morning Star about eight days after the inferior conjunction with the Sun. If we observed Venus through a telescope or powerful binoculars, we could see a continuously visible thin crescent—to the south at the approach and to the west as her morning phase begins. In mythology, Inanna and Ishtar were mourned as lost for three days, which astronomically is the time we lose sight of Venus. To the naked eye, during this phase the brilliant light of Venus is not visible. Venus is also lost in the sunbeams during her full phase—at the superior conjunction, when she lies on the far side of the Sun. Her cycles are symmetrical because the planet's orbit is the most circular of those of the planets in our solar system.

From our Earthbound vantage point, the superior conjunction—Full Venus—does not look like an opposition. However, figure 7 shows how Venus travels around the Sun. At the Full Venus, Venus is on the opposite side of the Sun. At the New Venus, Venus "hides" behind the Earth.

Quick Cycle Review

There are two main cycles. The first is the waxing and waning 584-day synodic cycle, which equates to about eighteen months and is split into two nine-month phases. There are five phase cycles, which create the five eight-year cycles. Five phase cycles equate to 2,920 days, which equal eight solar rotations (8 x 365 = 2,920 days). Thus, each one of the five synodic periods within the complete cycle repeats at 2,920-day intervals, which is eight Earth years. The numbers five, eight, and thirteen are intrinsically connected to the Venus cycles and repeated in the pentagram shape used to illustrate her cycle.

In addition, every forty years Venus and the Sun repeat the same relationship to each other. Transiting Venus is conjunct natal Venus, and the transiting Sun is conjunct the natal Sun. In Venus-Sun terms, life begins anew at age forty. We are amidst our mid-life crisis, which we typically assign to Uranus and Neptune. Women seem to come into their own at this age, often finding a sense of liberation. They feel free to be true to themselves.

The Pentagram: A Venus Cycle Symbol

Drawing that rose image (shown in the previous chapter) into our astrological wheel to plot the Venus cycle is challenging. When we plot the series of five Venus synodic points

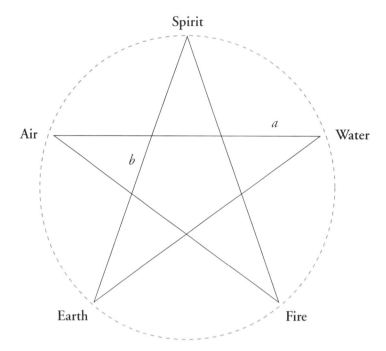

Figure 8—Pentagram with the Elements

and join these together around the 360° astrological wheel, the shape of the pentagram emerges. The pentacle—a five-pointed star—is formed by five straight lines connecting the vertices of a pentagon and enclosing another pentagon (five sides and five interior angles formed in the middle of the image) in the completed figure known as the pentagram. Before we look at the Venus points on this image, let us explore the symbolism built into the five-pointed star and its significance for understanding Venus and her cycle.

In figure 8, the top point of the pentagram is linked to spirit, and the elements of water, fire, earth, and air follow in clockwise order. If we were to place a person inside the pentagram, the head would represent spirit, the arms air and water, and the feet earth and fire. This is perfectly depicted in the drawing by Leonardo da Vinci of the Vitruvian Man inside the pentagram. The outstretched arms of a perfectly proportioned man—from tip to tip—equal his height.

The line identified as *b* in figure 8 is in golden proportion with any line marked *a*—1.61803. In the diagram, *b* represents the distance from one cycle point to the next, and line *a* is the cut-off section, i.e., the leg of each of the five triangles branching off the encompassed pentagon.

It is actually a revelation to note in how many instances the pentagram symbol is used. The symbol for the Venus cycle is an ancient symbol of protection. It is associated with Wicca and magic and is used in rituals. The order of invocation is from spirit (all there is and the divine) to fire (inspiration and spiritual cleansing) to air (intelligence, divine knowledge) to water (emotions, intuition, and surrender) down to earth (stability and physical endurance). The circle of protection flows in a clockwise direction. The banishing pattern flows down from spirit to earth and continues in a counterclockwise direction.[1]

Here is a short list of the symbolism of the pentagram:

- Gnostics—the Blazing Star, also known as Venus the Morning Star
- Freemasons—the 72° compass is the emblem of virtue and duty
- The women's division of the Masons is known as Eastern Star—the daughters of Venus
- In ancient and medieval philosophy, spirit was the purest element above earth, air, fire, and water, the foundation of life
- Hebrews—the symbol of Truth
- Pythagoras—the emblem of perfection
- Druids—the symbol of Godhead
- Egypt—knowledge and the Underground Womb
- Pagan Celts—the goddess the Morrigan
- Early Christians—five wounds of Christ
- Sir Gawain's glyph—symbolizing the five knightly virtues of generosity, courtesy, chastity, chivalry, and piety
- In advertisements—shows that a service or product is subject to a seniors discount
- Mormon temples—inverted form used
- According to Knight and Lomas, the pentagram sitting on two legs is the symbol for summer and the inverted one for winter[2]

The Number Five

The number five is significant in terms of Venus. Venus rules the five senses collectively, and the sense of taste, touch, and smell are closely associated with Taurus. Biblical scholars talk about the number five and its association with imperfection, lack, insignificance, and weakness when related to man. On the other hand, it is considered a number of completeness of the world. We have five senses, five fingers, and five toes, and we taste five flavors. We have five major organs (heart, lungs, stomach, liver, and kidneys). We have five basic virtues, and there are five special blessings: longevity, health, wealth, vitality, and natural death.

Here is a brief list of the main symbolism of the number five:

- In numerology, five represents an unbalanced state leading to change
- The square root of five is the initiate, Man in his fallen state
- Phi = ½ (1 + √5) = 1.618
- Medieval masons considered the number five and the pentagram symbols of deep wisdom
- Five stages or initiations in life
- Five wounds of Christ
- Five pillars in the Muslim faith

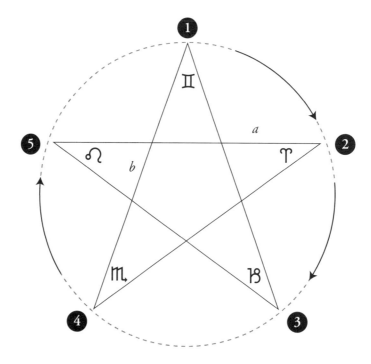

Figure 9—Venus Phase Cycle Pentagram

- Five of ten virgins were wise and five foolish, in the biblical reference
- The five wisdoms of Christ—incarnation, suffering (passion), ascension, stepping into heaven, and the second coming
- The path of five in life—longing for unity with god, faith, hope, humility and love; five mirrors unity

The Venus Cycles

Venus Phase Cycle

Figure 9 shows the Morning and Evening Star phases of Venus as they move from one sign to the next. Each Venus phase lasts 292 days or approximately nine months. The distance between each New and Full Venus is 72°, or a quintile. Venus travels about 288° around the zodiac during each phase. The Gemini point at the apex is numbered as 1. Assuming that it is a New Venus, the following Full Venus will be in Aries. In this sequence, the Capricorn point represents a New Venus, Scorpio marks a Full Venus, and so on.

The phase cycle from New to Full runs in clockwise order. In the diagram, the sequence is Gemini, Aries, Capricorn, Scorpio, Leo, Gemini, etc.

With each of the Venus cycles, you will notice that there are five zodiac signs involved. The signs change slowly, at each of the five points. Each sign stays at a point for a century.

Venus Synodic Cycle

The zodiac signs in figure 10 are the ones where Venus forms her conjunctions with the Sun—both the inferior and the superior conjunctions. We could turn the image clockwise and mark any spot on it as number 1. Figure 10 shows the inferior conjunctions—New Venus—in sequence starting with June 2004, when it took place in Gemini. This numbered illustration uses the invocation pattern discussed earlier in this chapter. Number 1 is the spirit point, 2 is fire, 3 is air, 4 is water, and 5 is earth. Each conjunction occurs approximately 144° from the prior one. The order for the eight-year cycle is always the same: Gemini, Capricorn, Leo, Aries, Scorpio, Gemini, etc.

We can add or deduct eight years and follow the repetitive pattern. Looking at number 1 in 2012, we have the next New Venus in Gemini; the previous one was in 1996. In 2018 there is a New Venus in Scorpio, and in 2002 we also had one. Each point retrogrades back in the zodiac, and after a century, the sign involved is the prior one in the zodiac. In 2041, the New Venus in Aries (number 4) becomes New Venus in Pisces. Using the tables

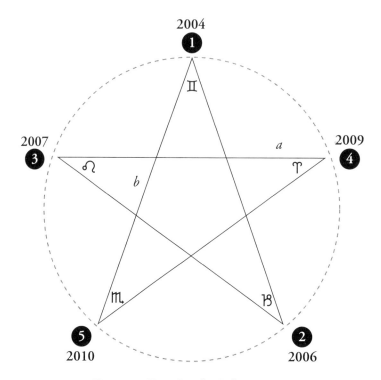

Figure 10—Venus Synodic Cycle Pentagram

or by simple addition and subtraction, it is easy to deduce that there will be a New Venus in Aries in 2009, 2017, 2025, and 2033. Further, we can travel back in time—after all, in order to divine the future we need to look at the past. There was a New Venus in Aries in 2001, 1993, 1985, 1977, 1969, 1961, 1953, 1945, and 1937. Prior to that, the point identified as number 4 in the diagram was in the late degrees of Taurus.

Sequence from New to Full Venus

Each New Venus is followed by a Full Venus four years later on about the same date and at almost the same degree. The conjunction point moves backward in the zodiac by about 2.5° in eight years. All of the Venus cycles have the same eighteen-month interval. Below is a table that lists the New Venus and Full Venus through the five signs, so that you can readily see two in each sign. The tables for each of the five cycles are in Appendix II: Venus Cycles Data.

New Venus	°	Sign	'	Full Venus	°	Sign	'
March 30, 2001	9	Aries	32	March 31, 2005	10	Aries	31
October 31, 2002	7	Scorpio	53	October 27, 2006	4	Scorpio	11
June 8, 2004	17	Gemini	53	June 9, 2008	18	Gemini	43
January 13, 2006	23	Capricorn	40	January 11, 2010	21	Capricorn	32
August 18, 2007	24	Leo	51	August 16, 2011	23	Leo	18
March 27, 2009	7	Aries	16	March 28, 2013	8	Aries	11
October 29, 2010	5	Scorpio	30	October 25, 2014	1	Scorpio	49
June 6, 2012	15	Gemini	45	June 6, 2016	16	Gemini	36
January 11, 2014	21	Capricorn	12	January 9, 2018	18	Capricorn	57
August 15, 2015	22	Leo	39	August 14, 2019	21	Leo	11

The Cycle of New and Full Venus in Gemini

The following table lists the New and Full Venus in Gemini over the course of one cycle. This information is also given in appendix II.

- The initial New or Full Venus in any cycle always begins at the end of the sign—around 29°.
- The cycle moves backward over the course of 104 years, shifting into the previous sign—in this case into Taurus.
- 13 x 8 years = 104 years; there are thirteen stations/conjunctions at eight-year intervals in each of the signs during which time 104 years pass.
- The New Venus, the Full Venus, and the retrograde and direct stations in each repeat at eight-year intervals, independently of each other.
- The average number of conjunctions and stations within a zodiac sign is thirteen, but only in terms of a 1,248-year full cycle through the zodiac.

New Venus	°	Sign	'	Full Venus	°	Sign	'
June 19, 1964	28	Gem	38	June 20, 1968	29	Gem	8
June 17, 1972	26	Gem	30	June 18, 1976	27	Gem	4
June 15, 1980	24	Gem	20	June 15, 1984	24	Gem	58
June 13, 1988	22	Gem	12	June 13, 1992	22	Gem	53
June 10, 1996	20	Gem	3	June 11, 2000	20	Gem	48

New Venus	°	Sign	'	Full Venus	°	Sign	'
June 8, 2004	17	Gem	53	June 9, 2008	18	Gem	43
June 6, 2012	15	Gem	45	June 6, 2016	16	Gem	36
June 3, 2020	13	Gem	36	June 4, 2024	14	Gem	30
June 1, 2028	11	Gem	26	June 2, 2032	12	Gem	24
May 30, 2036	9	Gem	16	May 31, 2040	10	Gem	16
May 27, 2044	7	Gem	7	May 28, 2048	8	Gem	9
May 25, 2052	4	Gem	57	May 26, 2056	6	Gem	2
May 23, 2060	2	Gem	46	May 24, 2064	3	Gem	54
May 20, 2068	0	Gem	37	May 21, 2072	1	Gem	45
May 18, 2076	28	Tau	26	May 19, 2080	29	Tau	37
May 16, 2084	26	Tau	16	May 17, 2088	27	Tau	28
May 13, 2092	24	Tau	5	May 15, 2096	25	Tau	17

Each of the cycles in Aries, Gemini, Leo, Scorpio, and Capricorn work the same way. The tables for all of the cycles with the conjunctions and stations are in appendix II.

Up to Four Months in a Zodiac Sign

Venus moves more slowly when she is about to turn retrograde, spending up to four months in one sign—the approximate duration of a Jupiter retrograde period. This extended Venus period has a long-lasting impact. Nothing is started in haste—we take time to process matters and issues through our value system and personal ethics. Remember also that the New Venus takes place while Venus is retrograde. During this time, we make new choices for ourselves that reflect our changed priorities. Venus at this point always asks the question "Do you love it and are you content?"

Nine Months: Morning Star and Evening Star

What is the significance of nine months, the duration of a Venus phase? The human pregnancy lasts nine months. It is also typically the time it takes us to build something to the point where it can sustain a life of its own. Many authors comment on birthing and gestating their book over a period of nine months. Many courses to learn a trade last nine months, and at nine-month intervals tradesmen return for shorter periods to add to their basic training. The school year used to run from September to May. Nowadays, escalated courses are offered in eighteen-month increments in a variety of fields. Many a romance lasts eighteen months and then withers away unless a commitment is made. The appropriate length of an engagement prior to a wedding used to be at least nine months. Old English wisdom states: "Winter them, Summer them, and Winter them again before marrying them."

We are all familiar with the waxing Moon and know to plan new activities for that phase. The waning phase is better for finishing projects and clearing the decks for the new phase. This ebb and flow is fast—two weeks each. The phases of Venus lend weightiness to Venus; she does more than just quickly transit our personal planets.

Precision in Timing

Venus loves precision! In January 1990, the New Venus took place at 28° 35' Capricorn (having spent three months in Capricorn). The area of our lives represented by the house with Capricorn on the cusp or intercepted in it was active for an extended period, giving us time to think about these issues at length. Note that the same degree of precision applies to each Venus cycle—stations and the conjunctions in each of the five signs. Refer to appendix II for detailed listings.

In January 1994, there was a Full Venus at 26°44' Capricorn. That is slightly less than 2° of separation. The subsequent New Venus in 1998 was at 26°07' Capricorn! Venus truly works each degree—and the planet we may have at that point—several times over the course of our lives. Let us look at a few stories that illustrate how Venus shapes our lives.

Some Venus Stories

An example case is a chart for Client A (figure 11). (Please note that all charts in this book were calculated using the Koch house system.) Client A has the North Node at 3° Aquarius in the ninth house and Mercury at 27° Cancer in the second house. Mercury rules houses two and five. In 1989, Venus stationed retrograde at 6° Aquarius on his North Node, applying by 3°. The New Venus about twenty days later was opposite his Mercury, which also rules his Moon-Jupiter conjunction in Gemini in the first house. On Christmas Day, the downstairs of his house was flooded by a foot of water as a water pipe burst, swamping major renovations that had just been completed for the holiday. The repairs were made with insurance monies paid promptly at the Venus direct station. At that time, he was also approached about relocating. In April, with the quarter phase of Venus, the details were worked out. That was about four months or close to 146 days later. On the day of the next Full Venus in November, he moved into his new house. In January 1994, with the Full

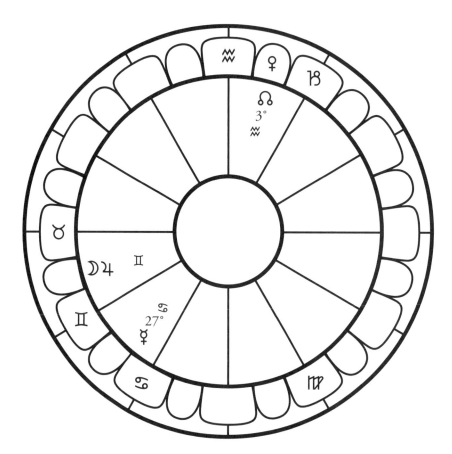

Figure 11—Partial Chart for Client A

Venus in Capricorn, he received a promotion with a significant pay increase and began studying for an additional university degree. In this example, we looked at both the short nine-month Venus phase cycle and the "culmination" of the Capricorn New Venus at the Capricorn Full Venus.

Until a natal planet is activated by the actual degree of a Venus station or conjunction, there are no significant events but merely themes, defined by the house.

Prince Charles of England was single until the age of thirty.[3] In 1980, Venus spent four months in Gemini, in his eleventh house, where his Uranus at 29° Gemini is situated. His seventh-house cusp has Aquarius on it, thus in modern terms his spouse is defined by

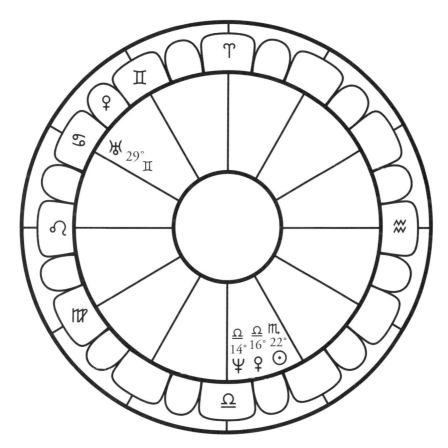

Figure 12—Partial Chart for Prince Charles of England

Uranus. At that time, a hunt for a suitable wife began. He met Princess Diana at the quarter Venus phase, in November of that year. In 1981, he was married following the Full Venus at 17° Aries in his tenth house, which happened opposite his Venus in Libra, which happens to be conjunct Neptune. Prince Charles has idealistic notions of love; it needs to be ethereal.

In late 1986, Venus stationed retrograde at 20°24' Scorpio, within 2° of Charles's Sun at 22°25' Scorpio, and in 1987, at the quarter phase, the media made his marriage situation world news. In 1996, the Venus retrograde period in Gemini once again danced on his Uranus, and at the quarter phase later in the year, the divorce settlement took place. In 2004, Venus once again spent four months on his natal Uranus—this time a powerful occultation. Shortly after the quarter phase, his marriage to Camilla Parker-Bowles was announced. The marriage took place on April 9, 2005, following the Full Venus at 10° Aries (March 31, 2005). Both Charles and Camilla have their Midheaven at close to 10° Aries.

Incidentally, by the time Venus formed an inferior conjunction at 10° Aries in the spring of 2001, Camilla Parker-Bowles had become an established figure in Charles's public life. The speculation is that Charles met Camilla in 1972. Back then, we once again had Venus in Gemini for about four months. Both Charles and Camilla have Uranus in a late degree of Gemini and both have Leo rising. Venus knows what the heart desires, and eventually we arrive there in eight-year increments with a highlight at four-year intervals. Charles's progressed Venus at 26° Sagittarius was opposite the New Venus of 2004. I guess he fought his private battle back then to make way for his second wedding.

Quick Review of Venus Cycles

While the Venus stations and conjunctions are separated in this book for clarity, it is important to note that they are integrally linked. The major factors to consider are that Venus spends up to four months in the zodiac sign of her retrogradation; the cycles may involve two zodiac signs as they shift, but even then, Venus spends about a hundred days in the zodiac sign of the New Venus. Look at those degrees in the tables in appendix II and study the illustrations in the following chapters to see when any of them were conjunct a natal planet in your own chart. Until the degrees become close and personal, the themes by house and sign play out but we may not act on them.

---

1. Note that there are many practices among Wiccans and others working with magic. The order of the Venus cycles is observed in this one.
2. Christopher Knight and Robert Lomas, *Uriel's Machine: The Ancient Origins of Science* (London: Arrow Books, 2000).
3. Unless otherwise noted, the data for famous individuals used in this book came from Lois Rodden's *AstroDatabank*, http://www.astrodatabank.com. Only charts with an accuracy rating of AA (from birth certificate) or A (from memory) were used.

CHAPTER FOUR

Venus Stations

Long after the final haunting note is played, its memory resonates. Scents and sounds evoke powerful memories in us—sometimes from such distant times.

Planetary stations are significant timers and show us where we need to pause to look at an issue in our lives with thorough scrutiny. A stationary Venus, like the final note of a piece of music, echoes long after the musician has struck the last chord, holding a powerful influence. Over time, we lose sight of what is truly important, thus we need to review life periodically. Stations stay active for a long time, leaving a permanent marker, which Venus will revisit time after time. These cycles help us understand timing with money, relationships, and our values.

Venus is about evaluation, reevaluation, and devaluation. Life changes subtly over time, and Venus offers us opportunities to reassess our altered value system. What we once appreciated and held in high regard no longer matters to us anymore. Just think back to what you dreamt your life would be four years ago, eight years ago, or in childhood. If you were offered that life today, you would probably deem it not important in the here and now.

Retrograde Phenomenon

On average, there are thirteen retrograde stations within a zodiac sign at eight-year intervals over a 104-year period. The retrograde period lasts approximately forty days.[1] Astronomically, the word retrograde refers to a planet appearing to move from east to west, counter to the direction in which most astronomical bodies move.[2] The word retrograde implies a retrospective. We have the luxury of looking back and reflecting and ultimately determining what our desires have become. These periods bring a reevaluation of the internal workings of our lives.

The Sun is the controller of the other planets and sets the boundaries of speed, distance, and declination. The Sun is the director of our personal play—the one with the script, the changes to the next scene, and all those connections to the other characters. Prior to the retrograde station, Venus reaches her farthest elongation from the Sun—47°—the most allowed by the Sun. Venus spends about thirteen days at the same degree during both retrograde and direct stations. When Venus is ahead of the Sun by approximately one sign, she stations retrograde.

Transiting Venus in regular motion might spend a day in aspect with a personal point. That is akin to a visitor stopping by for dinner, a pleasant evening, and perhaps a bouquet of roses. However, when a visitor comes to stay for an extended period, we learn more about that person while accommodating her needs and wishes. If we have a planet near a stationary degree, things get personal. The visitor—Venus—has an agenda, related to the planet, its house, and the house it rules.

In a Loop

The motion of a retrograde planet, in its travel around the Sun, forms a loop in the sky (figure 13). The loop is coming toward us, and we feel rather than see it coming. The phenomenon of the loop could be likened to observing a train moving at a different speed from a moving vehicle driving adjacent to the train. It feels disorienting. We have learned what happens when our computer goes into a loop, executing a command we do not recall giving it. The computer refuses to launch a new task; it is finishing what it thinks we asked it to do. The only solution is to reboot. That, in astrological terms, is similar to the New Venus, which begins the movement out of the loop.

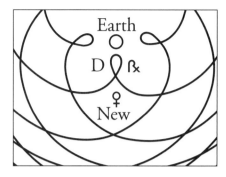

Figure 13—New Venus and the Retrograde Loop

A Venus retrograde period begins when she is at her most fragile, apparently motionless, having almost completed her 584-day journey. Venus is not in a position to receive the light of the Sun; rather, she is on her own "in the dark" and about to venture into that loop. During this period, Venus spends extra time in a sign or two. As Venus moves progressively slower as she approaches her retrograde station, we typically have the four months she spends in the sign of her retrogradation plus an extra month in the prior zodiac sign—a total of five months. Her travels around the Sun paint the rose pattern, part of which is that loop. On either side of the closure of the loop, we have a station—one retrograde and the other direct.

Let us think of the life cycle of an actual rose. We have a bud, like the New Venus, which slowly unfolds into a blooming rose at Full Venus. By the time we get to the retrograde station, we have a dried rose, which holds the memory of its beauty, preserved through eternity. A dried rose is beautiful but fragile, the color a faded version of her original splendor—"old rose." We used to assign that color to aging beauty.

Up Close and Personal

The speed of the planets is significant. Fast motion yields faster results or faster action. We typically consider retrograde planets as inwardly oriented with limited outward action. Retrograde planets act in more personal ways, and we have attached the moniker of karma to them. When something or someone is moving slowly, we have plenty of time to see and experience it. Venus zips around the far side of the Sun at a speed of 1°15' per day, i.e., close to the superior conjunction, or Full Venus. Her average motion through a zodiac sign is twenty-five days. Around the date the direct station occurs, Venus moves approximately four to nine minutes per day and gains speed relatively quickly.

When Venus is stationing retrograde, she is closest to Earth, i.e., at perigee. She is also at her aphelion point, which means farthest from the Sun. Once again, when something is in close proximity—right under our nose—it becomes more difficult not to acknowledge it.

We have five zodiac signs involved with the stations. Venus stations retrograde in a sign about thirty-two years before the inferior conjunction, or New Venus cycle, moves into the same sign. Many of us do find our thirty-third birthday significant. Our present history still shows that at age thirty-three many, such as Jesus, choose to end their lives with passion for life spent. What is it we say? "Only the good die young." Venus stations and conjunctions repeat every eight years in the same sign. The same degree is repeated a minimum of five times with the stations and conjunctions. Precision is her middle name!

The Statistics

Based on the 584-day cycle and the forty-day retrograde period, Venus is found retrograde 6.8 percent of the time. Five percent of individuals listed in AstroDatabank have Venus retrograde.[3] My personal database puts this number at slightly over 7 percent. Out of 14,161 charts with a Rodden accuracy rating better than B, 219 individuals were born with Venus stationary retrograde and 236 with Venus stationary direct—that is just over 1.5 percent. It is rare to be born with a retrograde Venus. Thus it logically follows that the individuals with Venus with a unique "condition" have equally unique issues to work on in their lives. We will look at Venus retrograde in chapter 9.

Phase Change

Venus approaches the Earth as an Evening Star and departs as a Morning Star—the phase is changing. Venus appears bigger and brighter as an Evening Star due to our perspective of her—she is closer to the Earth. As the Wise Woman or Sophia of Wisdom, she is less focused on the material side of life and more on relationships. With the retrograde periods we take a journey into the underworld—our subconscious or unconscious mind—encountering hidden or forgotten truths about ourselves. Whenever we talk about Venus, we are invoking something in our hearts. Venus is about what the heart desires. Her question during the retrograde periods is "Do you love it?"

A child born to older parents—often called the Evening Star—tends to get the best that the parents have to offer, as the parents' needs are now secondary to those of the child.

This seems to imply that the child is somehow older and wiser, or that we are. *The Evening Star Venus puts the needs and desires of others first.* The changing of phase is an interesting phenomenon in a progressed chart. Suddenly, if your progressed Venus becomes a Morning Star, you learn to put your own needs first. Conversely, if your Venus becomes an Evening Star—meaning you were born close to Full Venus—you learn to consider the needs of others as a higher priority than your own. Vedic astrology considers a retrograde planet stronger than a direct one. This is the case with the stations, which take place on either side of the conjunctions discussed in the next chapter. The retrograde Venus is definitely more personal, as it is closer to us.

In order to illustrate astrological concepts, we tend to paint verbal caricatures. One phrase to describe Venus as a Morning Star is "Diamonds are a girl's best friend." Diamonds, the ultimate precious stones, are said to last forever. Diamonds are associated with Aries. The woman who sang those famous words onscreen was Marilyn Monroe, who had her Venus in Aries. Love may not last, but diamonds, something of material value that can be exchanged for cold, hard cash, are akin to insurance. *During the Morning Star Venus phase, we tend to go after what we feel we are worth.*

Retrograde Stations

Retrograde stations leave a lasting impression, particularly when a planet or point in the natal chart is at that degree. Figure 14, once again in the shape of the pentagram, shows the retrograde stations of Venus, which in chronological order take place in Aquarius, Virgo, Aries, Scorpio, and Gemini. The signs change at each point over a century but the pattern remains. You will notice that the direct stations take place in a different sign. Venus travels back about 16° during her retrograde motion. Once again, please remember that the New Venus discussed in the next chapter takes place between the two stations.

Figure 14 shows three dates and degrees for each point on the pentagram. You may wish to calculate when Venus reached any personal points, and think back, or refer to the tables in appendix II. You can go back in time without consulting an ephemeris. Simply add 2.5°, deduct eight years, and add two (sometimes three) days. For example, look at the Aries point. There was a Venus retrograde station at 20° Aries on March 11, 1993, which was preceded by a station at 22.5° Aries on March 13, 1985. You can see that one of the stations or conjunctions (discussed in the next chapter) has touched or will touch your personal chart.

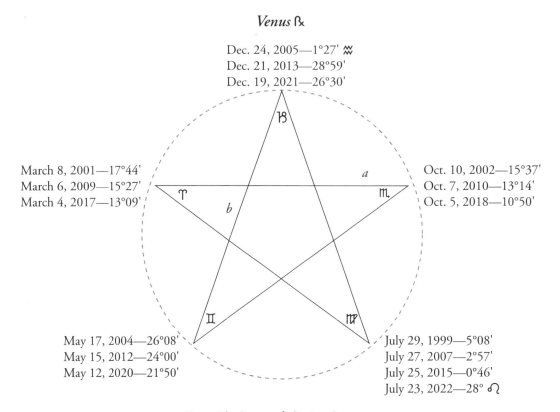

Figure 14—Retrograde Stations Pentagram

The diagram shows that the retrograde stations in Virgo are about to shift into Leo in 2022. The first New Venus in Leo, in our times, took place in August 1991. The first Full Venus in Leo took place in August 1987, and by now, the direct stations take place in Leo—and have since 1991. The baby boomers born at the end of the cycle have Pluto in late Leo, and in the first few decades of the twenty-first century will have ample opportunities to accomplish their personal mission in terms of the collective. The energy is there to invoke. The verb invoke has some interesting synonyms: bring into play, call upon, pray to, appeal to, to demand, to petition, to plead, and to ask. Thoughts can and do manifest—we have it in our power to create a different kind of reality and even a future for ourselves and our society.

During the period in which the stations occur, a span of about half a sign—16°—becomes sensitized. When these stationary points fall within 1–3° of your personal planets and points, the associated events and shifts in priorities become pronounced. We learn to pause to reevaluate our priorities and reassess what we truly value, and at the point when Venus moves ahead, we can see light at the end of the tunnel. We can move ahead knowing that we are truly worthy and deserve more.

Hollywood Stories

Charlize Theron, the stunning, versatile South African actress, was born with Venus retrograde in Virgo (August 7, 1975, in Benoni, South Africa[4]). Her public persona bears a distinct resemblance to the refined, elegant Ingrid Bergman, who also had her Venus in Virgo. On the day of Charlize's birth, retrograde Venus at 11°41' Virgo was square Neptune at 9° Sagittarius.

In 1991, Venus stationed retrograde at 7°19' Virgo within 4° of her natal Venus. That year her progressed Venus was at 6°21' Virgo retrograde. The sixteen-year-old won a local modeling contest. Four years later, she had her first film role—no dialogue but three seconds of screen time. That year, 1995, she had a New Venus by progression at 4° Virgo. In August 1999, *The Astronaut's Wife* was released with Venus retrograde—the station was at 5° Virgo. Charlize gained new roles and appeared in five well-received movies in 2000.

As an aside, the movie *Splash*—the retelling of the Little Mermaid story—left a lasting impression on Charlize as an eight-year-old. She knew in her heart that she could have played the part better, and was jealous of Darryl Hannah (Venus in Capricorn), who played the part next to Tom Hanks (Venus retrograde in Gemini). Perhaps Charlize's tune is to define who she is, perfecting her personal role, through acting on the silver screen.

Tom Hanks, who played the "Prince" to the Mermaid in *Splash*, was born with Venus retrograde at 22°50' Gemini on the Midheaven (at 20°42' Gemini). In 1980, Venus spent close to four months in Gemini; the retrograde station was at 2°35' Cancer, within a 3° orb of his MC ruler, Mercury, at 5°47' Cancer and conjunct his progressed Uranus. Hanks landed a part on the TV show *Bosom Buddies*, in which he appeared on screen in women's clothing. The benevolent Full Venus took place in June of 1984 on his Venus on the Midheaven, the year *Splash* was released. In 1988, Venus once again spent about four months in his tenth house within 10° of his Midheaven; he scored big, with *Big*. The New Venus was at 22°12' Gemini! He has won three Oscars for his acting—as an actor he is in a category all his own. Might he contemplate complete retirement in 2020, when the retrograde station of Venus takes place right on that Venus/Midheaven midpoint?

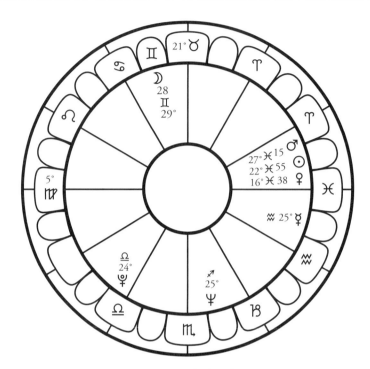

Figure 15—Partial Chart for Peter

Real-Life Stories

A young man we will call Peter decided that completing high school was not for him and dropped out of school in late spring of 1996. Venus stationed retrograde at 28°18' Gemini, applying by one degree to his natal Moon at 29° Gemini in the tenth house. His Moon is square Mars at 27°15' Pisces. Peter worked at a variety of dead-end jobs for eight years. In 2004, Venus once again stationed retrograde, at 26°08' Gemini. Peter returned to school, obtaining the first level of a certificate for a trade. He completed his training in record time, in seven months rather than the nine allotted. By the time he graduated, he had a job waiting.

There was a Full Venus in March 2005, the culmination of the nine-month phase cycle. He quit the job! His reasoning was that he deserved better treatment, better pay, etc.—he was definitely worth more. In early April 2005, he came for a consultation. He had applied for a position that offered a future and continuous paid training, and had already had three interviews. He was anxious to know if he would be offered the position. Stations leave an imprint that lasts. All it took was transiting Moon to reach that degree of Gemini, and a week later he

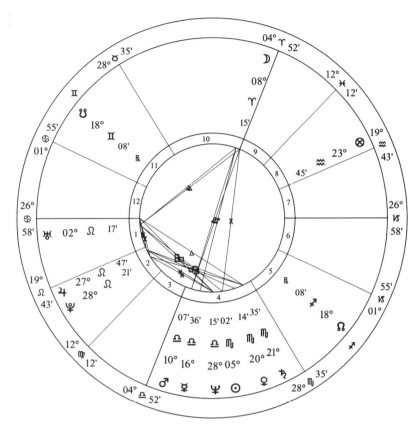

Figure 16—Chart for Bill Gates
October 28, 1955 / 10:00 p.m. PST
Seattle, WA / Koch Houses

got the position with more perks than he had envisioned in his wildest dreams. He had spent a long time deciding what he was worth, thus he negotiated with confidence.

Bill Gates,[5] the founder of Microsoft, has Venus conjunct Saturn at 20–21° Scorpio in his fourth house (using Koch houses). In October 1986, Venus stationed retrograde at 20° Scorpio, the year his company went public. A day after his birthday in 1998, the Full Venus was conjunct his Sun at 5° Scorpio in the fourth house, and the U.S. Federal Government launched a formal lawsuit claiming Microsoft to be a monopoly. In 1999, Venus stationed retrograde in his second house, and stayed in his money house for four months. The New Venus took place on his Jupiter-Pluto conjunction at 27–28° Leo; he made a pledge to donate one billion dollars to education.

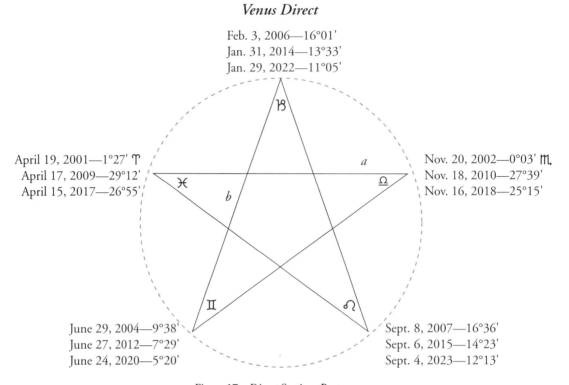

Figure 17—Direct Stations Pentagram

I was born about five hours after Bill Gates. In October 1986, I first learned about our impending move to the United States from Finland (Venus-Saturn in the tenth house). Around the time of my birthday in 1998, I was in a brand-new job, which paid more than the market average, and had extra time to pursue astrology (Sun in the ninth house). In 1999, I gave my first lecture at a conference—on Venus, I might add—and met some incredible astrologers who have been instrumental in the creation of this book (Jupiter-Pluto in the eighth house). As an aside, Bill Gates began his computer empire from a garage in 1975, the same year I began working on mainframe computers.

Direct Stations

The direct station follows the retrograde station in about forty days and about 16° earlier in the zodiac. The New Venus falls in the middle about twenty days and 8° after the retrograde station. Venus stations direct when she is one sign behind the Sun in the order of the zodiac—her cycle is symmetrical. As of this writing, the direct stations also take place in Aries, Gemini, Leo, Libra, and Capricorn. The last time any of these points were in Libra was back in 1903, quickly illustrating that a century just flew by. For review, the top point in figure 17 is number 1 and features the glyph for Capricorn. Point number 2 is in Leo and number 3 is Pisces. This point is about to shift into Pisces in 2009; the final direct station in Aries took place in 2001. Each sign gets to spend about a hundred years in the same position. The zodiac turns backward at each point in the course of 12 x 104 years, which equals 1,248 years.

This diagram also shows three dates and zodiac degrees for each of the five points of the pentagram. Once again, if you have planets at these degrees, the cycles of Venus "play" into the scheme of things in your life. We can quickly calculate back in time by adding 2.5° to arrive at the preceding station, and two days later in the calendar eight years ago. We can go forward by reducing the degree by 2.5°. Further, we add eight years and deduct two days.

In the short term, anything that was denied in terms of money or value during the retrograde period is reinstated at the direct station. It is our inner permission to move forward toward acquiring what we value. The forty days of subtle persuasion by Venus have left a lasting impression. While Venus is retrograde, monies and things owed are put on hold, and that theme ends when Venus stations direct. Venus retrograde does not seem to deny, but only delay. The timing element of Venus became an important tool when my neighbors asked me when the insurance adjuster would finalize payments for claims processed during the retrograde period. The direct station date worked very effectively. Before we look at a few examples of people with a strong or well-placed Venus who have a knack for passionate persuasion, let us look at a few Venus concepts in relation to the stations.

Fertility

Venus is linked to fertility. Her blessings brought abundance to the land and its people. Both the waxing and waning Venus phases number 292 days. Twenty days of each cycle mark the days when Venus is retrograde, leaving 272 fertile days. Pregnancies where the first trimester falls during the retrograde period are fragile. Miscarriages are a part of life—devastating for a time. Unplanned pregnancies, likewise, often happen during the Venus retrograde period. If we think of Lilith, the one Venus sent away in the story of Inanna, we might see a connection. Women used to wear amulets to ward off Inanna's influence. The poem of Inanna talks about the womb being ready to receive the seed. Lilith is associated with stealing those "seedlings."

Back in 2004, a young woman devastated by an unplanned pregnancy could not decide what to do. Venus stationed retrograde in her fifth house, where she has her North Node at 15° Gemini, which is opposite a natal Venus-Uranus conjunction at 11° Sagittarius. It had been a one-night fling while Venus was retrograde; surely she was not going to be a mother. Just prior to the direct station at 9°38' Gemini, she miscarried. With the Full Venus at 10°26' Aries in March 2005—her natal Moon is at 11° Aries—she moved out of town for a year to make a change in her lifestyle.

Another young female client conceived an unplanned child at the same time in 2004. When the Full Venus culminated in Aries on her Sun, she left her marriage. The father is suing for custody; the young mother just wants to play. A more upbeat story: A client, who has Mars at 25° Gemini in the fifth house, was desperately trying to get pregnant while Venus was retrograde in Gemini in 2004. She conceived after Venus stationed direct and returned to the degree of her Mars. Venus is a wonderful timer, which at times may also bring us what we did not realize we wanted.

Unlike fairy tales, real-life stories about Venus periods do not always conclude with winning an Oscar or making quick transitions. We are often faced with those nine-month transitions from something that "died" to something that we gestate. We develop an idea in our mind, study it for substance, and ponder how feasible our brainchild is. A seed for life, both figuratively and literally, needs nourishment in order to be birthed. Somewhere along the way, we pause to evaluate. Venus does give us a hundred years of each theme so that we can see who we are when no one is watching us. Have we earned or do we deserve our status, gifts, or blessings, or are we completely unworthy of rescue? It is our personal quest for inner wisdom and our own question to answer.

Note that individuals born with the retrograde station will by progression experience a change in about forty years. We have many tools for progressing the chart; the discussion in this book is limited to secondary progressions (taking one day to represent each year of life). When a retrograde planet turns direct by progression, we typically have a stunning change in our lives. It may be a time when we fall in love, or finally decide to marry or have a child. When her Venus retrograde in Scorpio stationed direct at age seventeen, a client moved to a new country that year to begin a different life. Once Venus moves direct by progression, there tends to be newfound freedom to try other things.

Marriage

One of the most popular marriage times in Europe happens to be in late June, on Saint Baptiste Day. However, in 1956, 1964, 1972, 1980, 1988, 1996, etc., Venus was retrograde in Gemini around that date. Do these marriages last? Some do and some do not. It depends on why the relationship was consummated at that time—to legalize a sound relationship or to marry for all the wrong reasons. Venus is about what the heart desires, not logic or tradition. Oftentimes, romances that begin with Venus retrograde tend not to last; people fall in love with love, not with the other person.

Individuals Born into the Venus Cycle

If we are born into any of the Venus cycles, that question "Do you love it, and are you passionate about it?" haunts us throughout life. You are drawn into the cycles if you have planets in the signs in which Venus stations or forms conjunctions. Venus in Capricorn teaches us to learn to always love being the responsible one, and live in accordance with societal rules. Venus in Scorpio might coerce us to go ever deeper into our inner world to find our true essence. With Venus in Virgo, we might find new and improved ways to continuously perfect facets of our personality and skills, as our flaws appear bigger than life. During the months when Venus is in Gemini, we learn to love ideas—the more the better. We become more inquisitive and communicative and learn to appreciate variety, the ebb and flow of life with its delightful changes. When Venus spends months in Aries, we may discover our independent streak, fight to keep our own identity, or take bold initiatives in ventures and romance that we had not dared to try before.

Individuals with Venus Stationary Retrograde

Famous individuals with Venus stationary retrograde include Muhammad Ali (boxer), Tom Berger (American novelist), Danny Bonaduce (*The Partridge Family* actor), Bjorn Borg (Swedish tennis player), Pat Buchanan (American politician), George Bush Sr. (former U.S. president), Frank Capra (Italian movie director), Jeff Green (American astrologer), Julio Iglesias (South American singer), Juan Perón (Argentinean president), Oral Roberts (American evangelist), Lloyd Robertson (Canadian broadcaster), Gianni Versace (Italian fashion designer), and Elsie Wheeler (American psychic behind the Sabian symbols).

The stories of these famous individuals born with Venus stationary retrograde reflect the Venusian dynamic profoundly. The images of the life work of these individuals linger on.

The impact of the idealistic, sentimental movies directed by Frank Capra still stirs heartfelt imagery. George Bush Sr., whose Venus is retrograde in Cancer, returned to the White House as the father of the president in 2000. The Full Venus in the summer of 2000 was conjunct his Sun. In 2004, Venus spent four months around the elder Bush's Sun, and he got his second shadow term behind the seat of power. The names listed above bring in imagery of scrappers with longevity—"Take me away kicking and screaming; I am not done!"

Jeff Green, with Venus retrograde in Scorpio on his Ascendant, wrote two books on Pluto, which delved into the depth of the human psyche—and thus launched evolutionary astrology. He retired with the long Venus in Aries cycle in the spring of 2001. The Evolutionary Astrology Network continues with his students teaching his life work. Elsie Wheeler's words live on, and these are reinterpreted in the many books written about the symbolism of the degrees.

"I am a lover not a fighter. I am a poet . . ." Muhammad Ali took pride in being a wordsmith, and chose not to pick up arms, becoming a famous conscientious objector. He would rather be imprisoned than kill another—rather a contrast to George Bush Sr. who took the U.S. to war to protect his people. We have choices in how we use the energies. Venus as the Blazing Morning Star is linked to warfare, yet all of the mythology seems to imply that she was the protector, not the aggressor. The swastika and the pentagram alike have graced battleships and planes—to protect rather than symbolize aggression. However, the five-pointed image is also about power, as in Inanna's emblems of crown, jewels, and measuring rod. The appearance of the Blazing Morning Star was a signal for going to battle with victorious results. Venus does represent what we own and rightfully belongs to us from our point of view.

Individuals with Venus Stationary Direct

At the other end of the cycle, we have people born with Venus stationing direct. There is a lot of stored energy in a stationary direct planet; the quality represented by it is strong. Tony Blair (British prime minister), George Carlin (American comedian), Deepak Chopra (Indian-American author and inspirational healer in holistic health), Jean Chrétien (Canadian prime minister), George Clooney (American actor), Heidi Fleiss (notorious Hollywood madam),

George V (British king), Mata Hari (Dutch spy), Michelangelo (Italian Renaissance artist), Julia Parker (British astrologer), Alex Trebek (Canadian game show host of *Jeopardy*), and Yoshihito (Japanese emperor).

It may have struck a chord that George Bush Sr. is in the previous list, and two other heads of state for Britain and Canada are listed here. It may strike another note to think about how the listed individuals either knew how to turn a persuasive phrase or learned how to do so. Chrétien won no international favors because he, like Muhammad Ali, did not want to engage in hostility, and sent no troops to Iraq.

Deepak Chopra is an inspirational speaker on spiritual issues. He has Venus stationary direct at 2° Sagittarius. The topics for his sold-out talks and popular books include "peace is the way," "seduction of spirit," and "magical mind and magical body." An excerpt promoting his book *The Higher Self* sums up his message: "Dr. Chopra merges modern science with spirituality to demonstrate how verifiable scientific evidence closely supports ancient metaphysical traditions—and how applying this knowledge can impact all areas of your life. He believes that you can align the energy of your physical body with the energy of the universe, and that by doing this you tap into an infinite reservoir of intelligence. This intelligence is the 'higher self.'"[6] Incidentally, when the first of his numerous books was published, Pluto was transiting conjunct his Venus in the ninth house. The topics alone show us what the message of a stationary Venus can be—the seduction of spirit.

Quick Review of Venus Stations

- Venus slows down steadily as she approaches the Earth, moving about 30' per day for several weeks prior to her station; around her Full phase she moves 1°15' per day, which equates to twenty-five days per zodiac sign. Around the retrograde period, her speed is approximately 21–24' minutes per day in reverse motion.
- Venus stations retrograde at 584-day (18-month) intervals, as an Evening Star.
- The successive direct stations of Venus as a Morning Star also repeat at eighteen-month intervals. New Venus takes place exactly halfway between the stations.
- For thirteen days around her retrograde station, Venus stays at the same degree. As an example, on August 6, 1975, Venus stationed retrograde at 11°43' Virgo, and was at

that degree from July 31 to August 12—accentuating the contemplative nature of the Evening Star.
- Venus remains retrograde for approximately forty days and forty nights—remember that biblical reference when Moses wandered in the desert, and the temptation of Jesus. Note that Venus retrograde by progression lasts for approximately forty years.
- Venus spends twenty days retrograde both as an Evening Star and as a Morning Star.
- *Venus spends up to four months in the sign of her retrogradation.* The period Venus spends in the sign of retrogradation, prior to the retrograde station, is typically referred to as the shadow period.
- When the station occurs early in the sign, two signs of the zodiac share that four-month period.
- Venus moves backward approximately 16° in those forty days.

---

1. The period during which Venus is retrograde varies from forty to forty-seven days.
2. Venus naturally rotates in retrograde direction. If we were to watch a sunset on Earth from Venus, the Sun would set in the east.
3. *AstroDatabank*, http://www.astrodatabank.com.
4. Source: *The Free Dictionary*, http://encyclopedia.thefreedictionary.com/Charlize+Theron.
5. Source: *AstroDatabank*, http://www.astrodatabank.com, Rodden accuracy rating of A—quoted by Bill Gates.
6. "The Deepak Chopra Online Store," *People Success*, http://www.peoplesuccess.com/chopra.htm.

CHAPTER FIVE

The Cycle of Harmony

Sacred geometry
Where movement is poetry
Visions of you and me forever

Oh, let me wheel—let me spin
Let it take me again
Turning me into light

The two stanzas are from the *Dark Waltz* by Hayley Westenra, a young New Zealand vocalist, on her 2004 CD *Pure*.[1] Does it not sound like pure Venus? Love songs tend to feature "I will be with you again . . . or forever"—that longing is innate in human nature. However, the words also serve as a reminder that the essence of Venus continually finds ways to stay in our consciousness. The Canadian band Tea Party recorded "Inanna" on their 1995 CD *The Edges of Twilight*.[2] Incidentally, their website features images of celestial spheres and sacred geometrical images. In ancient Sumer and in Babylon, we had hymns; nowadays we have contemporary songs. Our stories are more commonly told in song and on the silver or TV screen.

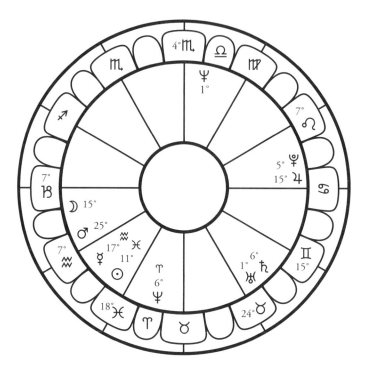

Figure 18—Chart for Julie

The Venus waxing and waning cycle, and her association with love and longing, suits the next story to perfection. Not all are this perfect in astrological precision. Julie, who has been an astrologer for decades, has her Venus at 6° Aries opposite Neptune at 1° Libra. Venus is sextile her Uranus-Saturn conjunction in early degrees of Gemini, and Neptune is in trine aspect to it. In addition, her Venus is square her Capricorn Ascendant and rules her fourth house. Libra is intercepted in her ninth house. In 1985, Venus was in Aries from early February to early June. Venus stationed retrograde at 22°18' Aries (March 13, 1985) square her Mars at 25° Capricorn in the first house. The New Venus (April 3, 1985) was at 14° Aries, square her natal Moon-Mars opposition to Jupiter. She met a younger man; the direct station (April 25, 1985) was at 6°00' Aries conjunct her natal Venus (her Venus is at 6°01' Aries). In June, they moved in together. On April 4, 1989, the Full Venus at 15°10' Aries was in square aspect to her Moon (15°) and Mars (25°) conjunction in Capricorn—her Moon is partile opposite her Jupiter. The younger man, who had shared her life for the past four years, proposed. Rather than leap up in joy, she woke up one morning simply knowing that she wanted something more out of life. She wanted stability and social, material, and financial benefits. What suited her needs in the spring of 1985 was not what her heart desired in 1989. As in the story of Peter and his career in the previous chapter, Julie decided she wanted and deserved more.

She spent the next nine months dating and watched every imaginable movie with her best friend. All the while, she yearned for a love from sixteen years earlier—the one that got away. In the midst of the inner turmoil, she even spent time on a psychiatrist's couch describing how all the men she met failed to measure up. Venus was in Capricorn from November 6 to December 10, 1989, and January 17 to March 3, 1990. The methodical Venus in Capricorn reached her Moon-Mars conjunction on the day she attended a Rotary Club function in late November of 1989. There, lo and behold, she met her lover from sixteen years earlier. It was the first New Venus in Capricorn in her lifetime—at 28°35' Capricorn! The following autumn, the pair were wed. Today she sees her Venus-Neptune opposition as having always been under the spell of someone—not seeing others for who they truly were. The Capricorn Full Venus in 2002 was conjunct her well-connected Mars and brought her significant recognition as an astrologer.

Neptune has an interesting effect. We need only to see a commercial for an allergy medication to understand the nature of this planet. It can deprive us of clarity; we may live in an altered state of reality. We will look at Venus in aspect in chapter 7.

Awakening

The eight-year cycles of Venus work to bring us in harmony with dormant desires, talents, and agreements we have made at a higher level. When we fall in love, we feel that with every cell of our body. When we search for a new home, the one we choose is the one that makes us feel we are at home. There is an old wives' tale about everything being for the best, even when we lose something we thought we loved. A phrase from years ago lingers in my memory: losing a job is nature's way of saying that you were in the wrong job.

A New Venus marks new beginnings, but as with any beginning, something else tends to end—the two go together. Venus is the most systematic in her motion. Eclipses work through degrees, as does Venus. We typically have four eclipses a year. These take place in pairs, and six months later, there are two more close to those degrees—a solar and lunar one. Then eighteen years later, we have a repeat performance. However, Venus does it with precision, persistence, and frequent repetition, never stepping out of rhythm.

Inferior Conjunction

As a review, each New Venus cycle begins with the inferior conjunction, or heliacal rising. Heliacal is a term we use to talk about a planet setting or rising with the Sun—*Helios* in Greek. Venus takes her sweet time at eighteen-month intervals, keeping us focused on about a sixteen-degree stretch of the zodiac. We tend to remember what we did for one third of the year—Venus does spend a long time, up to four months, in the sign of her retrogradation. Venus approaches us as the Evening Star, having completed her cyclical dance around the Sun.

During the retrograde phase, Venus moves more slowly than the Sun. Venus is attracting the Sun at this point of her cycle, waiting for the Sun to catch up to her. The new cycle demands a new pact, beginning, or covenant. My working title for this book since inception was *The Covenants of Venus*. Covenants means agreements, legally binding contracts, and mutual promises, as well as a lawsuit for breach of agreement. Law, what is fair and just, as well as agreements, are all Venus concepts. This chapter is titled "The Cycle of Harmony"; however, it can indicate life-altering events that do not feel harmonious in the least. Interestingly, one of the synonyms for harmony is accord, which means to agree, or to have consensus. It is about what is best for us; we just do not know it yet. Each new cycle is a fresh start within a specific area of our lives, and when the degrees activate a planet at that degree, this cycle becomes personal for up to forty years.

Venus in our natal chart defines what we crave, what we are working on in order to acknowledge our innate worth. If we consider the concept of reincarnation, Venus would represent what we perceive our value to be and what we think we are worth, both in terms of material possessions (Taurus) and in relationships (Libra). To value something is to have a high regard for it. Watch how during the times when Venus is in Taurus, we tend to feel that we are worthy or deserving of things.

We tend to become disillusioned or dissatisfied with the area of our lives that Venus is transiting during her retrograde period; or, at the least, we question whether the issues by house, sign, and aspect are truly worth it. Venus tends to awaken dormant desires and tendencies within us. The story of Julie at the beginning of this chapter was an illustration of how this woman decided she wanted something different from her relationship. She then took her sweet time contemplating what she really wanted.

Venus wears the "clothing" of a zodiac sign, and lives in one area of our chart and has dominion over two houses. By contrast, the Sun—the director of our personal play—is in charge of only one house by rulership. For illustrative purposes, let us consider someone with Venus in Gemini in the first house. What would she have to offer? She would love to be seen as a vivacious, charming conversationalist with the gift of putting others at ease. She would desire to be carefree, and in return, she would soothe you with her words.

What would Venus in Gemini for months on end expect us to reevaluate? Are we being heard, do we get to engage in lively discussions, and are our daily lives stimulating?

In classical astrology, aspects were defined as dexter or sinister; that is, approaching from the right or from the left. A simpler way to look at it is that the dexter planet is waxing, and the sinister one is waning. Our rules are that the faster planet applies the aspect. During the retrograde phase, Venus is slower than the Sun, so the Sun catches up to Venus. At the Full Venus, she moves faster than the Sun, racing to catch up to the Sun—she is the one doing the pursuing.

New Venus

The tables for the New and Full Venus stations and conjunctions are in appendix II. There are five cycles in each. Currently, New Venus takes place in Aries (1929–2033), Gemini (1964–2064), Leo (1991–2095), Scorpio (1930–2026), and Capricorn (1990–2077). The first New Venus in Aries took place in 1929; therefore, in 1921 we had the final New Venus in an early degree of Pisces. The Gemini series of New Venus began in 1964; thus in 1956 we had a New Venus in an early degree of Taurus as the cycle concluded. In 2076, we have a New Venus in a late degree of Taurus, as the Gemini New Venus cycle moves backward. In 1983, there was a New Venus at 1°25' Virgo. Likewise, in 1922 the New Venus took place in an early degree of Scorpio. In 1982, there was a New Venus at 1° Aquarius. Venus, astrologically speaking, becomes the Morning Star at the conjunction. She is still retrograde in motion, but a new cycle has begun.

In the New Venus series in figure 19, notice that the approximate 8° that Venus moves backward after the retrograde station can cause this conjunction to take place in the previous sign. For example, if you flip back to Figure 14: Retrograde Stations Pentagram in the

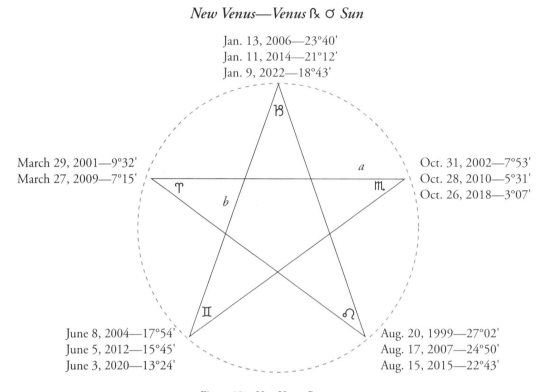

Figure 19—New Venus Pentagram

previous chapter, you will note that Venus stationed retrograde in Aquarius, but that the New Venus took place in Capricorn. The same applies to the New Venus in Leo, which follows the retrograde station in Virgo. Venus consistently retrogrades backward through the zodiac. The 2022 Venus retrograde station in Leo will be followed by a New Venus in Leo. The same applies to the shifting Aquarius-Capricorn point. Note what a wonderful ephemeris these four pentagrams provide. We simply need to remember that the difference is approximately 2.5° and about two days, added when looking back and deducted when looking ahead.

Three dates are given for each point in this inferior conjunction diagram except for the Aries point. We can do the math for this conjunction. If we deduct eight years from March 29, 2001, and add two days to the date, we arrive at March 31, 1993. Now we simply add

2.5° and we know the approximate position of the conjunction was at 12° Aries, give or take a few minutes (11°49' Aries). Going in the other direction from March 27, 2009, we add eight years, deduct two days and 2.5°, and arrive at March 25, 2017, at 4°45' Aries, give or take a few minutes (4°57' Aries). Incidentally, all of the calculations and ephemerides used in this book were calculated using Kepler 7.0 Software by Cosmic Patterns Software Inc.

That is the end of the mathematics in this book, other than a brief description of how to determine the exact phase of your Venus at the end of this chapter.

Venus begins to wax after making her conjunction with the Sun. The Moon is easier for us to see, and typically, while we find that the New Moon is good for starting things, in reality we seldom see much action until we see that crescent in the sky about 3.5 days later. The same seems to apply to Venus. Seen through a telescope, Venus only hides for about three days; however, we cannot see her without magnification until about fifty days after the New Venus. At that time, we tend to initiate action close to the crescent phase at seventy-three days. From New to New takes 584 days, which equals about eighty-four weeks or 19.45 months. From New to Full is 292 days, forty-two weeks, or 9.7 months. A quarter of the cycle is 146 days and one-eighth is seventy-three days.

In practice, separating the stations and the New Venus from one another does not work. These are separated in the book, to illustrate the repetitive degrees and patterns. We need to remember that the New Venus takes place while Venus is retrograde, which is not an ideal time to launch new action. This tends to be the time when we make new choices based on our changed values.

Full Venus

The Full Venus occurs approximately nine months after the New Venus. This cycle repeats the zodiac signs of Aries (1941–2037), Gemini (1968–2072), Leo (1987–2091), Scorpio (1926–2014), and Capricorn (1986–2073). Each cycle has the same eight-year interval with thirteen conjunctions or stations and a 104-year span. There are slight variances, which even out in the 1,248-year period, when the zodiac has rotated through at each point. These Full Venus degrees replicate the New Venus conjunction points. Thus, the New Venus at 17°53' Gemini in 2004 becomes a Full Venus at 18°39' Gemini in June

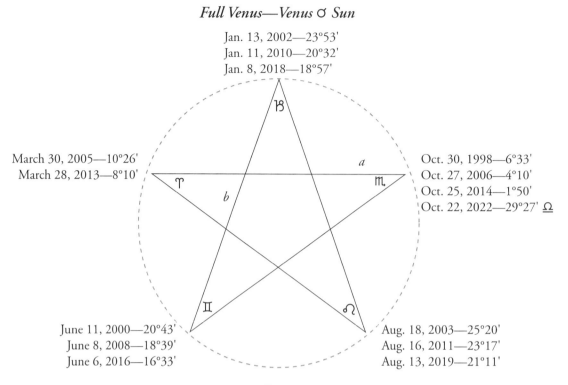

Figure 20—Full Venus Pentagram

2008. However, in between we have a Full Venus in Aries, a New Venus in Capricorn, a full Venus in Scorpio, and a New Venus in Leo.

The Full Venus point breezes by quickly and is easy to miss. However, during the weeks when Venus is in the sign in which she forms her conjunction with the Sun, the events and feelings in the lives of people seem to echo rewards for the choices made four years earlier with the New Venus. In addition, the sequence of events launched nine months prior comes to fruition. There are two cycles at play. The eight-year cycle repeats the degrees—from New to Full in four years, and back to New in another four years. This period seems to have eight one-year phase periods.

Typically, clients who book a consultation shortly after the Full Venus tend to comment on how they are still dealing with the issues that originated during the prior New

Venus period. A quick glance at the chart allows us to see if there is a personal planet within 8° of that New Venus. Both the New and Full Venus seem to equal an eclipse in strength. Technically, only the Venus occultations (discussed in chapter 12), which occur eight times per millennium, are akin to eclipses. Once a year, the transiting Sun returns to that position, Venus reaches the point in about nine months, and naturally, Mercury visits that degree about once a year. This appears to be a consistent timer for clients wanting to discuss the issues represented by the natal planet and houses activated, most commonly the house in which the New Venus period took place.

Secondly, we have the nine-month phase cycle between consecutive New and Full Venus points that are approximately 72° apart from each other. New Venus in Gemini is followed by a Full Venus in Aries, which is followed by a New Venus in Capricorn. We travel around the wheel in clockwise order. The sequence is in reverse order to the zodiac: Aries, Capricorn, Scorpio, Leo, Gemini, Aries, etc. If the last one was a Full Venus in Leo, then the next one is a New Venus in Gemini.

Orbs for Quintiles

In terms of the quintile series of aspects and their orbs, in the naturally forming pentagram, the quintile varies from 67° to 75°, and the biquintile from 136° to 152°, something to keep in mind when ascertaining aspect patterns. Quintiles are aspects often linked to the fifth harmonic, a system developed by astrologer John Addey and developed further by astrologer David Cochrane. The division of 360° by five gives us 72°. This aspect is most commonly assigned to our ability to link to the divine, and is considered a positive aspect. It holds little force, but rather offers opportunities we need to be ready to seize.

Venus Cycles in Your Chart

During the years that Venus is retrograde in the solar return chart, we seem to make new, well-contemplated choices for ourselves. The Venus periods at eighteen-month intervals become personal when our personal points are triggered by either the stations or the conjunctions. While mainly conjunctions are used as examples in this book, let us look at the story of the young female client discussed in the section "Fertility" in the previous chapter (the young mother who just wants to play) from the perspective of her husband. For the husband, the New Venus took place exactly opposite his natal Sun in Sagittarius in the second house. He has Libra rising; thus we consider his chart to be ruled by Venus. The direct station took place opposite his Venus. His wife left with the Full Venus in Aries, which incidentally formed a grand trine to his Mars in Leo trine Venus in Sagittarius (at the same his progressed Moon was conjunct his natal Venus). Sounds almost blissful, yet he was forced to sell the family home, left to pay massive credit debt, and even had to give the family pet away. He still dreams of reconciliation while devoting all of his free time to playing with his young sons.

Short Stories

I naturally tune in to the Venus degrees when doing a reading; thus, when I note a point in the chart in one of the signs in which Venus forms her stations and conjunctions, I ask the client what happened. Remember, Venus tends to bring us what we do not think we want. She awakens dormant desires and asks us to ponder whether we love what we have or are doing. The initial "hit" to a natal degree approaches from the right. Let us review a little: If a natal planet is at 26° Leo, the first retrograde station was perhaps 2° behind at 28°; eight years later, the aspect is exact. The first hit may set many things in motion; the second one is guaranteed to do so.

In 1970, Venus stationed retrograde on a client's Ascendant. The New Venus took place in her first house opposing her natal Uranus in the sixth house. A quarter of a century later, she still remembers what happened. She had a six-month-old baby in a house she loved, but her husband wanted to sell it and move his family to an investment property they also owned in order to buy a boat. She left him about four months later. The next New Venus took place in her eighth house; it made no aspects to her natal chart but merely spent four months in that house. She became depressed and suicidal, stating in hindsight that if it hadn't been for the children, she would not be here. In her heart, she knew that relationship was not what she wanted, nor was she prepared to cease living. Not all Venus stories are about love and harmony. A quarter of a century later, she is thriving and excited about every moment of her life.

During one of the first lectures I gave on Venus in the nineties, I heard numerous stories about the New Venus at 11° Aries in 1993. One woman had been forced to declare bankruptcy due to theft of merchandise at a fashion show, because someone else involved in the show had forgotten to ensure that the insurance policy was paid up. This in turn forced her to create a new career that was more spiritual in nature and in line with her personal convictions. Venus spent four months in her second house that spring.

The stories range from having to say goodbye to a dying relative and the repercussions of losing the family structure to discovering creative talents. Venus wants us to be comfortable with transitions and reminds us of the pacts we made before we were even born. Young adults are comfortable talking about past lives—it seems a given to them. Ultimately, when we lose a loved in death, we gain a "benefactor" with access to invisible realms.

New Venus on the Moon seems to be connected to buying that first home. The other element strongly aspected at the time we make those big decisions is Juno, the marriage asteroid. The New Venus marks the time when we simply know we desire something more. Most typically, the timing occurs when one of the inner planets, including the Sun and Moon, conjuncts the prior New Venus degree. Look through the listings in appendix II, find degrees that match yours, and note how you remember exactly what happened and what transpired because you acknowledged something deep within you.

Your Venus Phase

Determining whether you have a Morning or an Evening Star Venus in your chart is relatively easy. In chapter 2 we looked at how to see that at a glance. Place the Sun at the nine o'clock position. If Venus is below the horizon, she is an Evening Star; above it, a Morning Star.

Venus also has the same eight phases as the Moon: the New Venus, the waxing crescent, the first quarter, the gibbous Venus, the Full Venus, the disseminating Venus, the last quarter, and the balsamic Venus. These concepts were discussed by Dane Rudhyar in *The Lunation Cycle*, published in 1967. During the first phase, we are seeding and working instinctively; during the second phase, our new ideas begin to take root; during the third, we cultivate the new ideas; and during the fourth phase, we perfect. The full phase is the culmination, the sixth phase is about nurturing or nourishing others, the seventh phase is linked to disillusionment, and the eighth phase is linked to surrender. The Venus cycle is 584 days long; thus, dividing these into lots of eight gives us seventy-three days for each phase. At present, none of the astrology programs calculates the phase for us.

We can do simple addition to find the New or Full Venus that preceded our birth using the tables in appendix III. A Morning Star Venus is preceded by a New Venus, and an Evening Star Venus follows the Full Venus. Using Julie as an example (figure 18 earlier in this chapter), her Venus is at 6° Aries. It is behind the Sun in longitude, and thus an Evening Star. Therefore, her birth was preceded by a Full Venus. She was born on March 2, 1943. The preceding Full Venus was at 23° Scorpio on November 16, 1942. Using fingers or a spreadsheet, we can calculate that she was born 106 days after the Full Venus. Each phase is seventy-three days in duration; thus her Venus is a disseminating one.

As another example, Client K has her Sun at 21° Pisces and Venus at 29° Aries. The Sun would rise before Venus; thus her Venus is also an Evening Star. The Full Venus that preceded her birth was on June 24, 1952, at 3° Cancer. That was 261 days before her birth on March 12, 1953. That would make her Venus balsamic.

Prince Charles, whose chart (figure 12) was discussed in the previous chapter, has Venus at 16° Libra and the Sun at 22° Scorpio. He was born during the Morning Star phase. Therefore, a New Venus preceded his birth on June 24, 1948. Deducting that time from his birth date of November 14, 1948, we can conclude that he was born 143 days after New Venus. That would make him a first quarter Venus.

Another Technique to Determine Your Venus Phase

The quickest way to determine the phase of your Venus is to calculate a heliocentric chart.[3] All of the current software does this with a click of a button. The Full Venus is the one where she is opposite the Earth; at New Venus she is conjunct the Earth. The separation from the Earth gives you the phase relationship. Julie's heliocentric Venus is at 11° Taurus, which is 120° earlier in the zodiac than the Earth at 11° Virgo. Client K's Venus is at 2° Virgo approaching the Earth at 21° Virgo, and thus balsamic. Prince Charles's heliocentric chart shows Venus at 22° Leo and the Earth at 22° Taurus; that is a square all right—a fixed one at that. Venus is waxing toward full; therefore we know that this is the first square.

My Thoughts

After studying Venus rather intensely for sixteen years, and having spent hours gazing at the degrees, asking questions, and learning to understand Venus more intimately, I have found that we can invoke her energy. I love the Wiccan greeting "blessed be." I received a prayer from a government-licensed Wiccan practitioner who helped me immensely with copyediting. It goes like this: "I am a child of the goddess, I am well, I am happy, and I am loved. Abundance is mine, because as a child of the goddess, I am empowered to create miracles." No wonder Venus was worshipped in so many cultures. We are empowered to invoke that energy even in today's electronic society, by counting our blessings each day.

---

1. Part of the song is currently available at http://www.hayleywestenra.com.
2. http://www.teaparty.com.
3. Sun-centered. From this perspective, the planets are never retrograde and thus frequently found in another sign than the natal sign. My Venus is a waning Evening Star, about 36° ahead of the Sun; thus I have a Full Venus.

CHAPTER SIX

Where Does Venus Hold Court?

The astrological chart is our manuscript for life—we have the lead role. The planets are characters or archetypes with a role to play. The scenes are played on the twelve sets or stages we refer to as houses, places, or rooms.[1] The Ascendant is akin to the front door, and the Midheaven is our center stage—the doors are better known as cusps.[2] If we visualize a planet on a house cusp as being at the door greeting others into her home, it becomes easier to think of the planet's behavior. We may have stern Saturn guarding that cusp, with agreeable Venus as the resident; perhaps she is intercepted without ready access to the "door." The word intercept means to hinder people or objects from reaching the destination by stopping, delaying, or diverting them. This model can help us see how the characters might play out their scenes, and how they feel and behave under different circumstances.

In the natural wheel, the second house has earthy, pragmatic, physical, tangible Venus in Taurus holding the door. The social, accommodating, rational, fair Venus in Libra owns the key to the seventh house. These are the natural houses, homes, or sets of Venus. Further, in classical astrology, Venus is in her Joy in the fifth house.[3] It is helpful to group the houses to see how the planet feels differently in an angular house than in a cadent one.

Venus in each of the houses evokes a deep desire within us to give and receive, so that we can gain understanding into what we have to offer—what we have held in our hearts since birth or since the dawn of time. The Venus cycles bring some of our dormant desires to the surface, not all at once, but slowly over time. That forty-year stretch is so potent, perhaps she also holds the master key to our midlife crisis. We call that sporty car we just have to have a midlife-crisis car; we experience a lack of fulfillment regarding our inner desires and yearn to express our talents. We feel that there has to be more to life. We may make an apparently sudden decision to find ourselves and our innate talents. Venus does not always bring us what we would love to have, but rather what we did not know we truly wanted.

The house placement of Venus describes what we desire and love. Astrology is about adding layers upon layers to gain a fuller understanding of the position, placement, and aspects of the planets. Venus in the signs will be addressed in chapter 8. I have chosen this order because I look at the signs as the clothing the planets, or characters, wear in our life's play. For example, the color pink is linked to the heart chakra and budding love, and it is the color of the pin to promote support for breast cancer research. Barbara Cartland, a British romance writer, was considered the original "pink lady." Her fifth-house Venus is in Leo, and she wrote about love and dressed in pink. Mary Kay Cosmetics gives away pink Cadillacs to successful saleswomen. While I have been writing my book, many pink clothing items have found their way into my closet. I did not use to wear that color; it was supposed to be for baby girls only. The fashion industry promotes seasonal color trends to retailers about nine months ahead of time. Watch those pastel colors gain prominence when Venus is making pronounced statements in the sky. In 2004, with the occultation of Venus, soft green and various shades of pink replaced deep jewel colors. In essence, the fashion industry knows how to tempt us to part with our money while tapping into our heart's desire—even when we did not know we had a yearning.

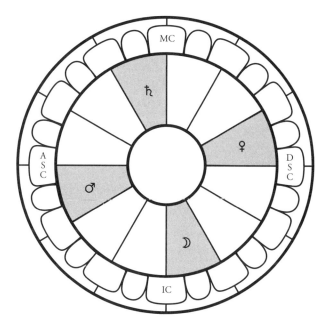

Figure 21—Angular Houses

Angular Houses: Our Rituals

Custom is the great guide of human life.
—David Hume

The cross and its symbolism were discussed in the introduction to this book. The angular houses directly relate to our personal issues and relationships, and to our obligations and actions in life. The horizontal line of the cross of cosmos represents the feminine principle and time—in the natural wheel, we are also looking at the masculine elements of fire and air. Planets that rule the pivotal angles in the natural wheel are in cardinal signs. The word cardinal means of foremost importance, principal, and originates with the Latin word meaning to hinge, connect, or link. In essence, these primary houses bind us to life. These four action-oriented segments of the chart are about our persona/image, family/roots, relationships/cooperation, and work/avocation. With the natural angles, fire desires recognition, earth achievement, air solutions, and water fulfillment.

The horizontal line from the Ascendant to the Descendant joins you and me. Linguistically, we always refer to others first, placing the self last. Indeed, it is considered impolite to place "me" first in a sentence that includes or refers to others in any way; Venus is diplomatic. The implication is not of natural willingness but rather that in order to accomplish her desire, Venus is willing to adapt. Harmony is priceless to her. In terms of astrology, we need to develop our sense of self, our persona, and our attitude before we can venture into relationships.

The qualities of the Ascendant by sign, ruling planet, and aspect show what we are cultivating and what wins approval. This point represents first impressions—unless planets have access to this angle through aspects, their energies are not readily incorporated into our public persona. The image we project is distinctly different with Venus in the first house than with, for instance, Pluto situated here. We tend to agree that the Venus person is more approachable. The first house and the Ascendant are our outlets for expression and personality, revealing our approach to life—what others see most readily, and what we may not.

Venus in the first house is charismatic, charming, and pleasing, and typically the person's physical appearance is striking if not beautiful. The Venus person tends to see herself mainly through the eyes of others while remaining blissfully unaware of these qualities. She laughs and cries with you and strives to be nice toward others. She yearns for compliments, yet is reluctant to receive them. She longs to be noticed, accepted, and appreciated. The following famous individuals have Venus in the first house within 5° of the Ascendant (the sign placement is listed inside the brackets): Claudia Cardinale (Italian film star, Taurus), Billy Crystal (American actor and comedian, Taurus), Donovan (Scottish musician, Gemini), Rush Limbaugh (American author and talk show host, Aquarius), Paul Newman (American actor and director, Capricorn), and Dr. Ruth Westheimer (flamboyant psychotherapist, Gemini).

When Venus stays in the first house for four months, we initially notice all kinds of flaws in our appearance and attitude and feel that we are not being appreciated. Eventually we decide that we need to make changes, and learn to appreciate ourselves. This can coincide with a parting of ways with people in our lives, as we deem that we are worthy of more. Trying to please more can also be a response, but ultimately that leaves us dissatisfied—we cannot buy love. The heart knows what the heart wants, and we can only fight it for so long.

The doorway, or cusp, at the IC, or midnight position, could be likened to the basement door—only used by family members. Venus and the Moon are the two feminine planets. The Moon represents our instinctive responses, memories, and emotions. The issues of her house, the fourth, include mother, home, family, and our subjective views about ourselves. In Vedic astrology, Venus is said to be well placed here, as it offers a sense of contentment. Venus in this angular house shows an amplified desire to have a family, home, and emotional security. These properties and qualities define our roots—where we came from and what we pass on to the next generation.

Feelings, likes, and dislikes are Venus attributes. Emotions and needs are concepts that belong to the Moon. Our emotional reactions are defined by the Moon, because she remembers how any sensation made us feel. There is a subtle difference between feelings and emotions aside from the definition that emotions are strong feelings. Those lunar responses are innate and instinctual. By contrast, Venus feelings are firmly based on our personal assessment about what we appreciate. We make our assessment based on a mental or physical sensation rather than relying on instinct.

Venus in the fourth house rejoices in a home and its location and contents. A colleague and friend with her Venus in Capricorn in the fourth has a delightful habit of gliding across the floors in her exquisite house, making "yum yum" sounds and smacking her lips to show how she much loves and adores everything in it. The matters of the house in which our Venus lives give us immense pleasure. Venus in the fourth may value privacy, but this is an angular, action-oriented house—we cannot hide. Indeed, love may come to your door, or through family connections and introductions.

The following is a list of famous people with Venus in the fourth: Louisa May Alcott (American author of *Little Women*, Capricorn), Prince Charles (British royalty, Libra), Julia Child (American chef and TV personality, Virgo), Phyllis Diller (American comedienne, Leo), Tarja Halonen (first female president of Finland, Scorpio), Britney Spears (American singer, Capricorn), Ringo Starr (British musician, Gemini), and Elizabeth Taylor (American actress, Aries).

Those long Venus periods bring the fourth-house issues to the forefront. We now yearn for a sense of belonging, to have a home, family, and emotional security. Reconnecting with family can be a strong desire. Many of us relocate far from our native lands; relocated Venus in the fourth house often signifies that we moved to a new location to fall in love and start a family. While most astrologers consider that starting a new relationship or marrying with Venus retrograde is not a good idea, people do, many very successfully. It may simply be an indication that there was a covenant or pact to finish off karma or obligations.

There are no guarantees that a relationship will last 'til death do us part. Relationships will be discussed in more detail in chapters 10 and 11.

The most elevated angular house, the tenth, is the location of the Sun at high noon. The astrological wheel is also a clock for measuring day and night. The concepts in these angular houses are classic and basic: the opposite of home is work, the opposite of mother is father, and the opposite of emotional is controlled. The various rulership books in astrology have alphabetical listings of words and their correlations to planet, sign, or house. A good book of synonyms and a dictionary go a long way to help us expand our astrological vocabulary. Let us look at the synonyms for the word classic: typical, archetypal, timeless, memorable, abiding, ageless, immortal, traditional, stylish, refined, formal, and conforming to established standards and principles. It is starting to sound like Saturn, is it not? He is the owner of this house, and defines and sets the rules! This house, apart from holding concepts of career, status, and reputation, also represents what we are expected to achieve.

The vertical axis of divinity originates at the IC and terminates at the MC, the apex of the chart, which also marks our headstone at the cemetery as our final mortal imprint. Naturally, we invoke from top to bottom, when we cross ourselves. This axis is the masculine one and links two feminine signs in the natural wheel—Cancer and Capricorn. Venus in this elevated house holds onto customs and has strong ethics and priorities about her role in society. She is a charming, enticing negotiator with ambition—goals, aspirations, aims, objectives, and a purpose. She pursues her desires methodically; in this house, time is measured in long periods.

Here are some famous individuals with Venus in the tenth house: Kathy Bates (American actress, Cancer), Boris Becker (German tennis player, Libra), Nicholas Culpeper (English astrologer and physician, Scorpio), Robert Denard (French mercenary, Taurus), Gary Hart (American politician, Capricorn), Henry VIII (British king, Gemini), Steve McQueen (American actor, Gemini), Sarah, Duchess of York (British royalty, Gemini), Vincent van Gogh (Dutch artist, Pisces), and Loretta Young (American actress, Aquarius).

When observing individuals born with cardinal angles, we can quickly note how dynamic and goal-oriented these people are. Life is planned and organized. It is about marketing and proper positioning or posturing; there are standards for all professions and lifestyles. Prominence and angularity does not determine the difference between notoriety and fame. When Venus spends time in the tenth house, especially when her stations and conjunctions aspect our natal planets, we deal with some or all of these issues.

Seventh House: Relationships

In the natural order, Venus and Mars are always opposite each other. Men and women are said to be as different as cats and dogs. In both instances, the pairings learn to accommodate or tolerate each other. Perhaps it would be more accurate to say that Venus does the accommodating, and Mars concludes that he might as well stop fighting it—what he can change about the essence of Venus is limited. From a spiritual perspective, here we are learning to integrate these naturally complementary forces within us through our relationships with others.

Venus owns the third angular house, namely the seventh house of relating and cooperation. Most of us set up a home and have relationships and work, while continuously developing and polishing our facade and personality. In the first house, we learn to put ourselves first. It is the house of *me*—my body, my attitude, how I present the rest of the "package" to others. In relationships, we attract someone to complement us. People we know and associate with are always represented by the seventh house. Each one of the houses has special people within it; this one describes all *others*. This is where we initially meet people, and through our values, needs, and personal ethics, we move them into other houses. From our marriage partner, to the lawyer whose services we may have to engage to dissolve that union, the seventh represents the area where we need others, and where the resulting environment places demands on us and our resources.

The primary task of this house is to learn the art of cooperation. Venus establishes the rules and guidelines for her tenants. What are the rules of relationships? When we are on our own, we can do as we please. We do not need to explain our actions and choices, or make excuses for our personal habits, nor justify our chosen environment. Within any relationship, we need to consider the feelings of others, make room for them, and learn to share. All of our planets are involved when we fall in love. In a man's chart, we look to Venus and the Moon to see what kind of woman he appreciates, and what kind of qualities he expects. In a woman's chart, we look to Mars and the Sun. Venus is our power of attraction; we send signals, consciously or not, and reel in what our heart desires.

The keyword desire is assigned to both Venus and Mars. The "want" part of desire belongs to Mars, and the "wishing and longing" to Venus. Men and women go after their desires differently. Venus is more about what we would like to receive, and Mars about what we are willing to offer and to take action to attain. In mythology, Venus and Mars had an ongoing, tempestuous love affair. Her allegiances lay elsewhere—she was married after all. Venus is not necessarily faithful; she has her own agenda and an enchanting demeanor, which aids her goals. Juno, the marriage asteroid, represents the wife, and will be discussed in chapter 13. For now, let's just remember that Juno is the jealous wife, who stays committed to the relationship regardless.

Should our natal Venus be in the seventh house, marriage, or having someone to share our lives with, would take on extra importance. This placement can lead to a lifelong search for a soul mate—sometimes with many relationships along the way. We will discuss Venus in marriage in chapter 11. Individuals with a seventh-house Venus are compassionate and acutely aware of the needs of others. It may not always appear so; a young man with a Pisces Venus in the seventh proudly sports his t-shirt that reads "Do I look like a "%#!&" people person!," yet he is as sensitive and caring as can be.

The following are famous individuals with Venus in the seventh house: Maya Angelou (American performing artist and writer, Pisces), Tyra Banks (American supermodel, Capricorn), Mike Bloomberg (Mayor of New York City, Aquarius), Jeb Bush (Governor of Florida, Pisces), Kurt Cobain (American musician, Pisces), Ralph Waldo Emerson (American author, Aries), and Heinrich Himmler (German head of the Gestapo, Leo).

Succedent Houses: Maintenance Power

Energy is the essence of life. Every day you decide how you're going to use it by knowing what you want and what it takes to reach that goal, and by maintaining focus.
—Oprah Winfrey

The word succedent refers to the houses that follow the angular ones. The word implies accomplishing something desired or intended through persistent, constant, stable effort. These powerhouses represent our values regarding money, time, love, children, creativity, death, reincarnation, commitment, friendships, acquaintances, and the significance of acquiring material possessions and intangible assets. We know how to hold on to things we think we own, and via the Venus cycles, we learn more of those lessons.

The second, fifth, eighth, and eleventh houses are the position of the Sun at the height of each season. At the height of spring, flowers are blooming, the leaves on the tree become mature, and the promise of summer is in the air. At the height of summer, the trees bear

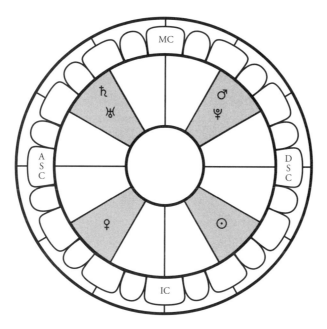

Figure 22—Succedent Houses

fruit and crops mature, and we have those sultry, hot days. At the depth of autumn, the land turns dormant in order to replenish, and in deep winter, we are resting and hibernating for the new cycle. Our personal values, and those of others, are reflected in the love we have to offer and affection returned. Our creativity and aspirations are realized through our personal efforts and with the support of those we love. In the angular houses we instigate, and in the succedent ones we expend sustained, persistent effort in order to maintain, and to have and to hold.

Some of the people most important to us are represented in these houses: those we choose to have in our lives. Their planets in your second house, and vice versa, are an indication that you share a common set of values; it is easy for you to agree. In the fifth house we have people whom we hold in our hearts. The sign and the ruling planet of the fifth house show us what we expect from those we love and what it takes for us to let them into our heart. In the eighth house we have those to whom we are committed, and in the eleventh we have friends and colleagues with whom we share interests. If you think of the important people in your life, you will likely discover that many of their planets occupy these four houses in your chart.

Collectively, these four houses are about our power and resourcefulness to sustain, acquire, and maintain. That encompasses our ability to provide for ourselves, put a roof over our heads, and have the means to keep on going. All of the succedent houses are linked to acquiring. The second house is of substance and sustenance; the other succedent houses also deal with money and resources. In the fifth house we do not appear to have a direct correlation to money, but let's explore for a moment—it is the house of children. We commonly spend the bulk of our income on supporting and providing for them. In traditional societies, children, in turn, are expected to provide for their parents in their old age. In that sense it still is a house of security and support. Like all financial projecting, it is also the house of risk, gambling, and taking chances. The eleventh house represents resources we accumulate through our career. In the eighth house we have joint resources as well as loans, debts, mortgages, and inheritances—and calculated risks and investments. In the eleventh house we have income and resources we bring in from career-type activity. Before we can leap into the other houses, we need to build a solid, secure foundation in the prime money house—the second.

Let us talk about Venus in her Joy in the fifth—that saying from classical astrology. A frequent statement by someone with Venus in the fifth is "I want to *play*" or "I want to have *fun*" or "You cannot *win* unless you *play*." We know that Lady Venus is a patron of the arts and a protector of women of all kinds. In classical astrology, this house is linked to the Soul. A primary role of Venus is to keep time and periodically remind us of things we had long since forgotten.

What delights us if we have Venus in the fifth house—our children, and creative offspring, the love of our life, and even gambling? Perhaps she shows our desire for applause, living life to the fullest, and seizing each moment to enjoy the richness of life. Metaphysical and spiritual teachers remind us that we need to learn to be happy and to play like children.

Here is a list of famous individuals with Venus in the fifth: Piercesare Baretti (Italian president of the Soccer Club of Florentina, Capricorn), Max Bill (Swiss painter, sculptor, and architect, Scorpio), Jeff Bridges (American actor, from large family, has three daughters, Capricorn), Geena Davis (American actress and MENSA member, three husbands and twin boys, Pisces), Princess Diana (British royalty, Taurus), George Harrison (British musician, devoted father, Pisces), and Steven Spielberg (American film director, seven children including adopted children, Scorpio).

Contemporary astrology assigns the eighth house to the soul. This succedent house deals with life and death, reincarnation, and the inevitable issues of life—from taxes and debt to the repressed issues in our lives; from depth psychology and the psyche, commitment, and sharing, to accumulating resources. In the eighth house we learn to eliminate issues and people from our lives that have ceased to support us. Relationships with a strong eighth-house involvement are considered karmic. Perhaps this implies that these feel fated and appear to resonate with the distant past. We seem unable to give up on these relationships; rather, we need to work continuously to make these viable. A garden thrives when we tend it and remove the weeds that suck the energy from the flowers we have chosen. Remember, the succedent house relationships are the ones we *choose* to have—they are not by chance.

In the eighth, we delve into everything in depth and with obsession, whether we want to make money grow, or grow our concepts and values to the extent that we can present them to the world. This list of people with Venus in the eighth seems to illustrate that concept rather well: Stephen Arroyo (American astrologer, Scorpio), Warren Avis (American pilot and founder of Avis Rent a Car, Pisces), Madame Billy (famed French madame, Taurus), George Blake (Dutch double agent, Sagittarius), Barbara Bush (American First Lady, Gemini), Kim Campbell (Canadian prime minister, Aquarius), Coco Chanel (French fashion designer and perfumer, Leo), Nicolaus Copernicus (Polish astronomer, Aries), Celine Dion (French-Canadian chanteuse who signed a hundred-million-dollar contract for four years of work in Las Vegas, just after fulfilling her deepest desire to "produce" a child, Pisces), Henry Ford Jr. (chairman of Ford and its founder's oldest son, Libra), and Billy Graham (American evangelist, Scorpio).

One interesting moniker given to the eleventh house is the place of the good spirit. What is the difference between the soul and the spirit? Spirit is character, strength, will, and courage of mind—and our spirit can be broken. Soul is linked to psyche, heart, essence, and spirit—and is eternal. In each of the succedent houses, our character, essence, strength, and spirit are tested, not through implicit action but our ability to sustain what we have materially and who we are morally. Temptation is the ultimate test of character—will we remain true to our essence?

In the eleventh house our earnings are based on the effort we expended in the worldly tenth house. Succedent does imply effort—nothing is free. This house is linked to dreams, wishes, and aspirations—we build our virtual castles in the air. We cultivate and work toward success. We have gifts in this house that we want birthed into the world. Influence, service, and the role of benefactor are concepts we associate with this house—the world is our friend. It is also about leaving a legacy. Just look at the following list of names of people with Venus in the eleventh, and the mark these individuals have made.

Harry Belafonte (American actor and singer, much humanitarian work, Aries), Tony Blair (British prime minister, Aries), Sir Jack Brabham (Australian race car pilot, knighted for service to motor sport, Aquarius), Harry Chapin (American songwriter, benefit concerts for world hunger, Sagittarius), Jodie Foster (American actress, owns her own film production company, single mother by choice, Scorpio), Llewellyn George (American astrologer, founded Llewellyn Publishing, Cancer) and Ariel Sharon (Israeli prime minister, Aquarius).

Second House: Value

In the natural wheel, the second house belongs to Venus. The concepts here are money, personal assets, values, self-worth, and possessions—time is an asset. We do claim that time is money. Taurus is assigned that possessive phrase "I have." Things that we value are to be found in this place as well, thus its association with priorities.

In older Hellenistic texts, this house is referred to as the gates of hell. "Money is the root of all evil with the power to change us." From a spiritual perspective, life here on earth can feel like hell, and many believe that there is no hell beyond—this earthly realm is it. If you recall, the introduction to this book mentions the attributes of Venus as the giver and taker of life.

We validate, nourish, withstand, and offer support based on our personal ethics, which is a Venus concept. Venus in the second house can be materialistic in the sense of desiring a stable source of resources. Our needs are unique—what is plenty for one might not be nearly enough for another. Other parts of the chart need to support any statement; one qualifier seldom says it all.

Now that we have emphatically noted the importance of money, let's look at other matters pertaining to this house of security. Priorities define what is important, more important, and most important, comparatively. We talk about personal resources—the means at our disposal. Our ability to produce is an asset that can be turned into cold, hard cash. The second house, by virtue of its ruler and the sign on the cusp, shows how we identify with material possessions. Physical senses, especially taste and touch, and talent, a productive skill we can bank on, belong to the second house. We need to learn to separate our self-worth from our monetary worth, rather than treat these two interchangeably. With an earth sign on this cusp or an earthy planet ruling this house, we tend to ensure that our financial picture is healthy, in order to function in the material world. A water sign on the cusp or a watery sign ruling this house might make the needs to "have and own" emotional ones. In the psychological hierarchy of needs, the main ones are food, shelter, and security. We have to be on a first-name basis with our second house to have a solid foundation to build our lives upon.

Venus in the second house takes delight in good food—preparing it, tasting every morsel, and enjoying how it looks. All of the succedent houses are strongly linked to the material side of life; however, the second house in the natural order belongs to Venus. She loves being here. The "collector" Venus, in the second house, loves holding on and accumulating more; her role here is to learn to do just that.

Famous individuals with Venus in the second house include Jane Austen (British author, Scorpio), Simone Beck (French chef, Cancer), Erma Bombeck (American journalist and humorist, Cancer), Johannes Brahms (German composer, Gemini), Nicolas Cage (American actor, Aquarius), Johnny Carson (American TV personality, Sagittarius), Francis Ford Coppola (American film director and producer, Pisces), Charles de Gaulle (French president, Sagittarius), Lee Dubin (won twenty-five million dollars in a lottery, Aquarius), Queen Elizabeth II (British royalty, Pisces) and Peter O'Toole (Irish actor, Cancer).

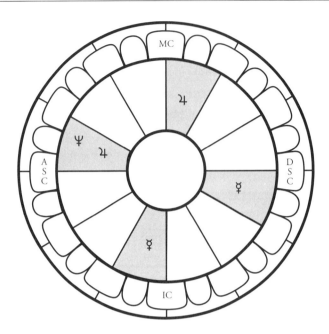

Figure 23—Cadent Houses

Cadent Houses: Knowledge

The union of the Word and the Mind produces that mystery which is called Life . . .
Learn deeply of the Mind and its mystery, for therein lies the secret of immortality.
   —The Divine Pymander

The cadent houses are about knowledge—the word and the mind. In the mercurial houses we collect and analyze information. In Jupiter's houses we seek to understand meaning by moving from the mechanics of "how" to "why."

The third and sixth are Mercury's houses—he defines the natural terms and conditions under which a planet can set up residence. The accomplishment of the third is "what I have learned and continue to learn." Venus in the third takes delight in information and insights through books, movies, and talking about it all. The sixth-house accomplishment is the skills and expertise acquired, through continual perfection. Venus in the sixth loves to create beautiful results, from the perfect bandage to a movie for the silver screen.

The ninth and twelfth belong to Jupiter, whose role encompasses sharing wisdom and teaching. In the classical order of planets, Jupiter and Mercury are always opposite one another. In contemporary astrology, we have assigned Neptune to the twelfth house. In the ninth house we seek higher knowledge and wisdom to broaden our perspective and search for understanding. In the twelfth we seek unity, and learn about faith and surrender, seeking ultimate insights into the human condition and all there is.

The quality of the third and sixth houses, the matters and issues falling in the domain of Mercury, are unique and different. The third house is where we accumulate basic knowledge and information required for developing skills. The sixth is where we apply that knowledge and develop the skills. Routine is doing something until it has been practiced to perfection—a skill. The inspiration found in the sixth house is watching or admiring your handiwork take form, be it a poem or bookcase. The fruits of our labor can and do manifest!

Becoming an expert at any craft requires a lot of third-house effort initially, which when brought into the sixth becomes skillful application. Only after we have acquired adequate knowledge and practical skills can we begin plying our craft. In order to become a master, we need to move to the next level. The how needs to be understood—our "holy grail" takes us into the realm of philosophy, theology, and the wonder of it all. The key phrase for the ninth house is "I see"; we learn to see, to understand, through contemplation and by listening to those wiser than ourselves. When the planets of others fall in our ninth house, we are in the presence of those who can teach us. The twelfth house is also a learning house. The knowledge we seek here is elusive; we cannot open a book to research these concepts. In this house we learn to trust our intuition without facts and proof. The analysis carried out in the sixth, when transported here, only holds a starting point, not the ultimate answers.

What would Venus in the third house find enticing? In addition to books, movies, dictionaries, and encyclopedias, she would love conversations with everyone in her daily life. She would know all her neighbors and the local gossip, and she would likely love the taste of words. Venus in this role would become involved in local groups or politics and maybe even set up a news empire. I have a midwife for birthing this book—a dear friend and professional astrologer, who offers continuous encouragement and valued insights. Janice has Venus in Aries in the third house and is a self-professed newshound, knowledgeable in too many subjects to list. She has penned romance novels, astrological columns, articles, and short stories. In addition, she managed international psychic fairs, and co-launched a successful 900-line for psychics and astrologers. She has a Buddhist frame of mind and practices Qigong. Along the way, she raised two children and had a successful fashion career—her ever-youthful image graced Canadian media for two decades. I will talk about all of the wonderful women who have helped with the process of writing this book in the Acknowledgments.

Famous individuals with Venus in the third house include: Dave Aaron (American UFO researcher, Libra), Robert Altman (American film producer, Aquarius), Alexandre Arnaux (French novelist, screenwriter, film critic, poet, playwright, and journalist, Aries), Simone de Beauvoir (French writer and existentialist, Aquarius), Agatha Christie (British author, Scorpio), Nat King Cole (American singer, Aries), Dalai Lama XIV (exiled Tibetan spiritual leader, Leo) and Betty Friedan (American feminist pioneer, Aries).

While the third-house Venus loves to share information, leaving you to decide what to do with it, the sixth-house Venus presents this as "researched" knowledge. Venus in the sixth enjoys perfecting the work and scrutinizing it from all angles. She strives to separate the noise from the pure sound. In many myths and fairy tales, the test assigned is to literally sort out and organize grains. Venus sent Psyche, her future daughter-in-law, to do this unaided to test that she was worthy of her son, Eros.

The list of famous people with a sixth-house Venus holds some premier names for giving the world a huge legacy of knowledge: Natalie Cole (American singer, Aquarius), Marlene Dietrich (German actress, Aquarius), Sigmund Freud (Czechoslovakian psychiatrist, Aries), Ernest Gallo (American vintner, Pisces), Robert Hand (American astrologer, researcher, and scholar, Sagittarius), Naomi Judd (American singer, Capricorn), Leonardo da Vinci (Italian painter and inventor, Taurus), and Shirley MacLaine (American actress, author, and dancer, Pisces).

The ninth house is most commonly linked to travel, higher education, law, and foreigners. This is the house of opinions, convictions, and teaching. Venus here loves gaining new insights and formulating ideas and always questions the information. Our Venus lessons in life might involve living overseas, being an eternal student, and studying philosophy, theosophy, astronomy, or any other similar expansive body of work. Venus in the ninth desires to understand, travel, and broaden her perspective. She is the promoter who loves to inspire and be inspired.

Thus far, understanding the truth of the Torah has been a 2,000-year-plus-long study by Jewish scholars. Each answer in the ninth house births a new question or quest. Venus in the ninth inspires us to seek the truth behind the facts or ideas. She would be delighted to unravel the "higher mysteries," by continually updating her personal ethics and values. All things foreign, new, and exotic would grab her fancy. We are here to become more than we ever anticipated—bigger, better, more knowledgeable through our increased understanding.

Here are a few notables with Venus in the ninth house: Howard H. Baker (American lawyer and senator, Capricorn), Lucille Ball (American comedienne, Desilu Studios, Virgo), Brigitte Bardot (French actress and animal rights activist, Virgo), David Bowie (British rock star and actor, Sagittarius), Sylvia Browne (notable American psychic, Scorpio), Deepak Chopra (Indian-American inspirational speaker, Sagittarius), Sean Connery (Scottish actor, Libra), Billie Jean King (American tennis player, Libra) and Sydney Omarr (American astrologer and journalist, Sagittarius).

Faith, fear, doubt, and contemplating the wonders of our inner world are twelfth-house concepts. In this, the house of the rising Sun, Venus delights in gaining understanding of these concepts and develops a highly unique set of values not based on logic, but rather on intuition, insight, dreams, and visions. We cannot learn everything from books, professors, or teachers; in order to truly know and understand, sometimes we need to sit alone and meditate on the essence of life.

This is a solitary placement of Venus, with a strong yearning for liberation from the material world. A twelfth-house Venus has a harder time learning to appreciate and accept herself; however, she is typically very kind and understanding of others. Facing reality can be challenging and threatening to our personal sense of security, especially about "who we are" and "what we believe." Winston Churchill talked about his personal demons, saying, "The black dog is back!" He would fade out of the limelight to deal with depression. Consequently, he became a competent landscape artist. In both stories and real life, the old queen often retired to the seclusion of a nunnery, away from the hustle and bustle of life, to contemplate the purpose of her life and the "beyond." Venus in the twelfth house is focused less on material luxury than on gaining strong spiritual insights. In this house of seclusion, those Venus questions resonate internally: "Who am I when no one sees me?" and "Do I love and accept myself just as I am?"

Individuals with Venus in the twelfth house include Ben Affleck (American actor, Libra), Jim Bakker (American TV evangelist, Aquarius), Lloyd Bridges (American actor, Pisces), Yul Brynner (Russian-American actor, Cancer), Tommaso Buscetta (Sicilian Mafioso, Cancer), Winston Churchill (British prime minister, Virgo), Leonard Cohen (Canadian author and songwriter, Virgo), Claude Debussy (French composer, Leo), Jimi Hendrix

(American rock musician and guitarist, Sagittarius), Mary Tyler Moore (American actress, Aquarius) and Winona Ryder (American actress, Scorpio).

---

1. In classical Latin texts, these are domiciles, and in Greece, the word was Topos = place. In Finland, these divisions are referred to as rooms.
2. The original meaning of the word cusp is simply a point.
3. Mercury is in his Joy in the first house, Mars in the sixth, Moon in the third, Sun in the ninth, Jupiter in the eleventh, and Saturn in the twelfth.

CHAPTER SEVEN

Venus in Aspect

"Here's looking at you, kid!" The famous quote from the movie *Casablanca* with Humphrey Bogart and Ingrid Bergman gets to the essence of aspects. The word aspect has its origins in the Latin word *aspectus*, which means to view, see, or look at.

This discussion on aspects is intended to provide tools for looking at the connections Venus makes with the planets rather than offer analysis on each variable. Classical astrology limits aspects to the Ptolemaic aspects—the conjunction, sextile, square, trine, and opposition. Major aspects represent outward action. The primary aspects describe the major characteristics, themes, and events. The wiring in our old house provides quite an interesting perspective. When we turn on the dishwasher, one of the three spotlights in a track light turns on and off until the dishwasher has finished its cycle—perhaps akin to a "loose" or minor aspect. We also have one switch that turns on the light in a bedroom while simultaneously shutting down a power outlet in another (this went unnoticed for about a decade)—perhaps akin to the subtle aspects, which are ours to discover? Harsh aspects require effort on our part; the easy ones can make us take something for granted.

Major aspects are easier to spot in an astrological chart. Squares occur within a modality (cardinal, fixed, and mutable), and trines within an element (fire, earth, air, and water). Even in the sky, these tend to be visual. The night sky in mid-October 2003 allowed us

Figure 24—Venus/Mars Parallel

to observe Mars and Venus equidistant from our perspective. In figure 24, Mars in Pisces is on the left and Venus in Scorpio on the right. We cannot superimpose a wheel against the real sky, but we can extend our arms out toward these two planets. Those arms lowered down are in the position we take to offer a hug. What a visual for a trine—a friendly greeting. The opposition would be to hold both arms stretched away from our bodies; we can perceive this as a balancing act, the best expression of this energy. It is a difficult pose to hold if supported by our shoulders, but more doable if we support the pose with our bodies. Arm movements are part of yoga postures; there is an easy way and a stress-causing way to perform them. The same applies to all aspects. It is stunning to watch planets align at the same height. This is what a parallel looks like. That particular parallel was partile several days after the trine aspect.

Venus Rulerships

Figures 25 and 26 illustrate how Venus looks at the other planets from her two rulership positions, Libra and Taurus. The natural wheel shows the planets in the natural houses and signs.[1] Venus in each sign is trine Mercury, opposite Mars, sextile Jupiter, trine Saturn in the same element, and square Saturn in the same modality. Venus in Libra, in a masculine sign, forms a sextile to the Sun and a square to the Moon. Venus in Taurus squares the Sun in Leo and is sextile the Moon. The greatest elongation between the Sun and Venus is less

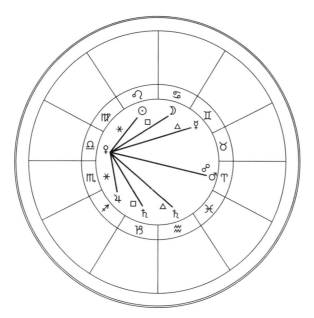

Figure 25—Venus in Libra Aspects

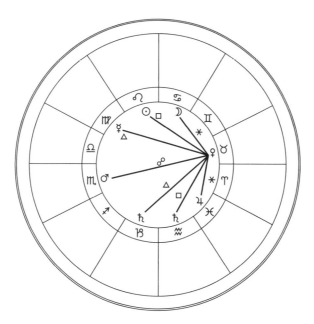

Figure 26—Venus in Taurus Aspects

than 48°; therefore, the only major aspect possible between the two is the conjunction. Venus and Mercury can form a sextile aspect—60°. These graphics feature only the seven planets known to classical astrology. In modern terms we would place Uranus in Aquarius, Neptune in Pisces, and Pluto in Scorpio.

Classical astrology assigns the feminine planets to the night sky and the masculine ones to the day sky. Venus in Libra gains strength in a day chart (diurnal chart); she has great rapport with the Controller, also known as the Sun. Contemporary astrology gives Uranus rulership over Aquarius. Venus in Taurus is sextile the Moon in Cancer. No wonder Venus rules Taurus by night and Libra by day.

Classical astrology also used to assign friends and enemies among the planets. Al Biruni states that Jupiter is friends with Venus by nature and asks for her friendship, while Saturn offers his friendship to Venus.[2] Jupiter rules Pisces where Venus is exalted, and Saturn is exalted in Libra, which Venus rules. There is natural rapport between these two planets and Venus.

Easy and Challenging Connections

It is all about having the right connections—how often do we hear that it is not what you know but who you know that matters? How well do we know others? With the easy aspects—the trine and sextile—we have a great rapport. With the trine, we get favors without asking, and with a sextile, we may have to put in a request. Planets in a square or opposition aspect make us work harder; the tune is adversarial, and we may need to give some in order to gain what we desire.

Conjunction

The word conjunction means union or combination. It punctuates, adds weight, gives strength and prominence, and can fall into either category of aspects. The planets are the actors in our life's play. Some share space and information freely, and others are rather territorial. Most of us have had to share our quarters with someone. If that space is large and we have plenty of elbow room, we cope fine. If we are confined to a sleeping bag in a tiny tent with a few other occupants, we tend to feel crowded, making nice or antagonistic gestures in an attempt to claim enough space for ourselves—the potential theme of a conjunction or stellium.

People behave differently as a couple than as individuals. When we observe a couple who have spent decades together, we notice that they often resemble each other physically. Perhaps the shy young woman has become more like the boisterous, outgoing man she married, and he has mellowed and become less outgoing. In a union, we learn from each other.

There is a strong reaction when a station or a conjunction occurs within about 3° of a natal or progressed planet. Around the New Venus, we have two stations; thus there are three zodiac points accentuated—the degrees of two stations and the New Venus degree. These degrees hold the note for years; however, it seems that if there is to be action, it will manifest within a year.

Square

In the square aspect, the elements are contradictory in nature—fire to water, earth to fire, air to earth, etc. Emotions can boil over; practical considerations can extinguish the passion, or logic counters innate security needs. The imagery of arm wrestling suits this aspect—the stronger planet wins.

How might Venus square Uranus function differently than Venus sextile Uranus? Uranus loves freedom and the ability to shock others. The square aspect would take the challenging road to achieving this, perhaps fighting every step of the way. The sextile would find an easy acceptance of the quality within and an easy outlet for expressing uniqueness, rebellion, and eccentricity. Any Uranus-Venus aspect describes a person or quality that is freedom loving, highly independent, and less likely to choose a relationship that inhibits their ability to do and be as they please. The role of Uranus is to shake up areas of our lives where we have stagnated and have not attempted to do anything unique, so that we can sing "I did it my way."

The square is an extremely productive aspect. Action requires friction—that is what makes machines operate. It can be an argumentative aspect, which allows us to test our limits and boundaries.

Opposition

We use the words compromise or disagreement to describe this aspect and typically link it to learning to accommodate and cooperate. It is the natural aspect between Venus and Mars, Mercury and Jupiter, the luminaries and Saturn—opposing views or stances. It is the relationship axis. In the natural wheel, this links together Mars and Venus. The themes of this axis were discussed in earlier chapters.

Sesquiquadrate and Quincunx

These two aspects fall into the minor category. A quick characterization for the first one is "Oops! I did not mean to do that." It seems to be present in surgery charts and when we fracture a bone or break a tooth, etc. The inconjunct, or quincunx (the glyph looks like a teeter-totter), is an adjustment aspect, continuously perfecting whatever the planets represent.

Novile, Decile, and Quintile

These aspects are referred to as harmonic aspects and are credited to Johannes Kepler. Based on geometric shapes and linked to the music of the spheres, they are considered to grant us opportunities and talents; the choice to take action is ours. The successive Venus stations and conjunctions of the phase cycle are separated by the quintile aspect (72°). The biquintile (144°) is the angular separation between successive Venus stations or conjunctions in the same zodiac sign—the eight-year cycle. The decile family of aspects includes the 36° aspect (also called the semiquintile), 72° (bidecile or quintile), and 108° (tridecile). The novile aspect is in increments of 40°; thus the trine would belong in this series. Michael Munkasey, an American astrologer and author, made an intriguing comment to me about the tridecile in a casual, private conversation in the late 1980s; he called it the goddess aspect.[4]

These aspects are subtle in nature. However, there is a connection, in terms of the electrical wiring mentioned earlier—we can flip a switch and turn this current on. The four aspects are linked to the unfolding of karma. They seem to be in great supply in friendship synastry and appear to negate some of the frictional major aspects. In a natal chart, these aspects seem to add nuances as well as alleviate the stressful aspects.

The theory on quintiles and deciles is that they are linked to the ability to tune in to higher wisdom, and once tapped, are like a blessing. Perhaps if we have a stern Saturn-Venus connection in our relationship sector or ruling it, and there is a nice quintile from Uranus or Neptune to Venus, we might learn to be quite sociable and even a tad more unpredictable. The novile aspect appears to be more active in the latter part of life, and allows us to integrate the two energies linked together by accepting ourselves, flaws and all.

However, in terms of forecasting what may be and even what has been, I work exclusively with the major aspects.

Applying and Separating Aspects

Aspects are either applying or separating from exact, adding extra hue to the aspect. An applying aspect is tenser; the energy is continuously building, with a sense of urgency and immediacy. The separating aspect, when activated by transit or progression, tends to be more internalized, as the skill, facet of character, etc., represented by the aspect is something already learned or experienced and not something that needs to be worked on, but rather something that has been earned as a birth right.

Declination

Figure 24 at the beginning of this chapter shows Mars and Venus at equal distance from the horizon. Parallels and contraparallels offer another connection between the planets. The Sun is in the northern declination from the spring to the autumnal equinox. The Sun's motion creates a perfect half-circle each six months, reaching 23° North 27' at the summer solstice.[3] During this time, the Sun is between 0° Aries and 30° Virgo. The second half of the circle is formed in the southern declination from the autumnal equinox to the spring equinox, and the winter solstice marks the time when the Sun stands still at 23° South 27'.

We measure these off the equator. Even on a sundial we see the Tropic of Cancer (north) and Tropic of Capricorn (south).[4] Declination is the measurement of angular distance above or below the equator along an hour circle.[5] Parallels are similar in nature to conjunctions, and contraparallels are similar to oppositions. There may be no aspect between planets that share the same declination. We might have a harsh aspect between two planets, yet an underlying parallel negates or diminishes the impact of the aspect.

Planets positioned in a Northern declination are active, masculine, day planets. The Southern ones are passive, feminine, night planets—black and white; in other words, duality. Color theory in computerized applications makes an interesting differentiation between the two non-colors—black and white—and the RGB spectrum—red, green, and blue—ignoring yellow as one of the three primary colors. The differentiation is that black and white represent values; in philosophical terms we often talk about life not being black and white (judgmental) but rather, shades of grey.

Venus changes declination when she is positioned in the first four degrees of Aries or Libra. Depending on the year, Venus can be found in the northern declination from early February to mid-November, but the stretch of tropical zodiac needs to be early Aries to early Libra—the equinoctial signs. Her pattern, when plotted on a graph, looks like the rhythm of the heartbeat. The average resting pulse during our adult years is seventy-two beats per minute. The one song with this tempo we may all recognize is "Oh, Christmas Tree," which is common in several countries. Perhaps the statement "It makes my heart sing" is more factual than we think.

Venus Out-of-Bounds

When planets venture beyond the highest declination the Sun reaches at the solstices, they no longer answer to the "Controller." Planets that line up at the same declination are similar to a conjunction in nature. Planets at a high declination exert a stronger influence. Think of it as sitting at the mountaintop, where you can see the big picture with more ease, as opposed to sitting under a tree at sea level, where only your immediate surroundings are readily apparent.

Venus only reaches that elevation when she is in Gemini or Cancer (north) or Sagittarius and Capricorn (south). A planet in the northern declination is considered stronger, more forceful, and more inclined to act than when found in the contemplative, southern portion of the sphere. The top part of the astrological wheel is the active, demonstrative, outgoing, masculine sphere of worldly accomplishment. The southern declination is the more private, personal, introspective, feminine, accommodating sphere. Venus can be found in the northern declination as early as the beginning of February and as late as mid-November.[6] Having a majority of planets in the southern declination can give us an internal sense of being grounded, whereby we are able to adjust to changes in life with more ease. If the bulk of the planets are in the northern declination, we may find it more disconcerting to make changes.

Venus out-of-bounds often occurs around Christmas time. Those seem to be the years when our chant is, "I am worth it; indeed, I am worth more." It is similar to a planet being unaspected, but only in relation to the Sun—so when it comes to spending or love, we simply want more, and are less likely to ask for approval. We simply go get what we desire. Incidentally, Venus ventures out-of-bounds when she is at about 12° Gemini or 12° Sagittarius—not every year. She returns within the bounds at the zodiac degree of 25° Capricorn on the day the Sun reaches 12° Sagittarius. When Venus is retrograde and out-of-bounds, the period lasts longer. As an example, Venus was out-of-bounds for two months in 2004.

Daily news during Venus out-of-bounds periods can be quite illuminating. In the final days of a recent one, there was an item in the newspaper about how people who had acquired new homes in the prior few weeks might not be able to afford them if they had opted for a no-money-down option. Common sense takes a back seat when Venus wants what she wants.

Mutual Reception

Planets are also in aspect through mutual reception; that is, when two planets are in each other's signs. For example, Venus in Aries and Mars in Taurus are in mutual reception by rulership.[7] Mutual reception is like an agreement between friends to share space and possessions and switch places. There is good rapport or affinity between the two energies. Also, look at planets and houses in this light when comparing two charts. If one person has Venus in Gemini and the other has Venus in the third house while there is no actual aspect, there is an affinity. Any two planets in mutual reception add a strong element to any chart.

Venus Aspects

Venus-Moon

The New Venus, conjunct a natal and even a progressed planet, seems to have the impetus of an eclipse. It works as a timer, but we can also make a highly educated analysis of the potential outcome by sign, house, and planetary connection. The Moon and the fourth house are connected to home and family. A young woman had Venus retrograde station on her Gemini Moon in June 2004. In 2005, when transiting Venus was briefly conjunct the Moon, she bought her first home. She bought it with her boyfriend, who had the 2004 New Venus on his natal Mercury, which rules the fourth house in his chart. His progressed Venus was at the same degree as the Venus retrograde station in 2004. They took possession when transiting Venus in Gemini reached his progressed Venus. In these two charts, the orb was less than a degree. The closer the aspect is to exact, the stronger its influence.

With a Venus-Moon aspect, we have an emotional need for financial security. We have a high emotional quotient, or emotional intelligence. It is easy for us to understand the feelings and emotions of others. Attachment to family, mother, and our roots is strong, and if we were denied that, we ensure that we create it. Possessions have sentimental value. This connection can endow us with a knack to generate the resources we most desire. The Venus-Moon connection might mean that we acquire money from our mother, but if that connection is in Taurus or in the money houses, we might have an innate talent for making money grow. We invoke what we desire with our Venus energy.

Venus-Mercury

On average, Mercury visits with Venus once every four months, and some years only once. The time lapse between two successive conjunctions in a zodiac sign can be up to eight years. The inner-planet patterns are infinitely unique. By knowing the Sun sign and one inner-planet aspect (or even just the signs the inner planets are in), we can quickly determine the year of birth.

Mercury is retrograde at the conjunction 36 percent of the time (178 out of 496); Venus is retrograde 26 percent of the time (128/496), and both planets are retrograde twelve times out of 496 conjunctions. The time span is from 1900 to 2100 (200 years). Both are direct 40 percent of the time (202/496), so actually it is rarer to have both planets direct at the conjunction.

Below is a table for Venus retrograde conjunct Mercury retrograde.[8] Note that in the first eighty-eight years of the twentieth century, there were three, and in the final decade, five—and in the twenty-first century, only four!

There is a societal factor at play here. The aspect between Venus and Mercury in the natural wheel is the trine—earth and air. Our desires and values are linked to our mind and speech. Half of these conjunctions involve Mercury retrograde.

Venus Rx conj. Mercury Rx, 1900–2100 (GMT)

Date	Time	Degree	Sign	Minutes
July 15, 1908	18:10	8°	Cancer	10'
June 30, 1948	19:52	29°	Gemini	09'
June 9, 1988	15:10	24°	Gemini	17'
February 5, 1990	1:06	21°	Capricorn	09'

Venus Rx conj. Mercury Rx, 1900–2100 (GMT) (continued)

August 11, 1991	9:24	5°	Virgo	24'
April 18, 1993	12:27	4°	Aries	05'
January 27, 1998	7:47	20°	Capricorn	20'
August 27, 1999	20:42	22°	Leo	42'
May 25, 2028	0:54	15°	Gemini	49'
November 1, 2034	23:59	22°	Libra	20'
May 11, 2068	12:31	5°	Gemini	58'
January 2, 2078	19:33	25°	Sagittarius	29'

Venus-Sun

With this aspect, self-worth and the ego are melded together. A rare conjunction is called the cazimi, which is when the planet is within 17 minutes of the Sun. Oprah Winfrey, the gorgeous American talk show host and billionaire who is still single at fifty, has Full Venus conjunct the Sun in Aquarius. Her photograph always graces the cover of her magazine *O*. Her logic is that she does not want to glamorize the glamorous. Glamour is absolutely Venus's domain. Oprah's struggle with her weight and appearance has been very public. She does a lot of good work to help people cope with life, donates huge amounts of money, and seems to work both sides of that Venus through the humanitarian sign of Aquarius.

American entrepreneur Leona Helmsley, nicknamed the Queen of Mean, amassed a fortune in real estate; her Venus-Sun cazimi conjunction is in Cancer. American film director Martin Scorsese has Sun cazimi his Evening Star Venus in Scorpio. This is not a common aspect, thus examples are scarce. The search through the databases turned up three famous names and three former co-workers—two of whom were charming and one a full-blown narcissist. All three examples have Venus as an Evening Star at the Full Venus.

When Venus and the Sun are not conjunct, we have a helpful separation between ego and our desires. We are not what we earn; we do not need to assess our worth by how others see us. When the two are in different signs, the separation is easier still.

Venus-Mars

In mythology, Venus and Mars had a passionate, stormy love affair. The natural connection between these two is the opposition, which represents both conflict and compromise. Venus is naturally the energy of cooperation, and Mars the energy associated with action and assertiveness. Once a relationship is established, we relocate the people into other locales. We have learned that opposites attract. However, if we place the opposite ends of magnets next to each other, they repel.

With this aspect our desires, wishes, and yearnings hook up with our will, determination, and motivation. Perhaps the easiest connection is when the two are in mutual reception. Venus in a Mars-ruled sign has easy access to the masculine. Women who have this in their chart "handle" men with ease, as they understand their motivation. Understanding other females may be more challenging. We know that the primary male mode of operation is to fix, hunt, and gather. The feminine mode is to support, acquiesce, and find an amicable, negotiated compromise. The role of a woman in our society still is to be the helpless one, to need the man to take care of her and to defer any major decisions to him. Remember, women gained the right to vote at the beginning of the twentieth century, and it is still not universal. "Men are logical and women are not" is still a perspective held by the majority. You cannot attack male logic with the same; women are expected to play out the emotional role. In her book *A Room of One's Own*, Virginia Woolf, with Venus quincunx Mars, wrote: "Women have served all these centuries as looking-glasses possessing the magic and delicious power of reflecting the figure of a man at twice its natural size."

Women with any Mars-Venus aspect, especially a harsh one, are not willing to play that game. They know how to look after themselves and do not intend to toe that stereotypical line. In contrast, women with the mutual reception have an innate knack of coyly playing the part. Venus gets her needs met and plays the part she feels is best to attain her goals.

Two famous individuals with a Venus-Mars conjunction are Mohandas Gandhi (in Scorpio) and Adolf Hitler (in Taurus). *Note, no single aspect tells the whole story!* Ed McMahon, co-host of *The Tonight Show*, had some stormy marriages with his Mars square Venus, and the American actor Bruce Willis also has the square aspect. John Wayne Bobbitt (Venus in Taurus opposite Mars in Scorpio) became a news item when his wife cut off his penis and threw it away.

The Venus-Mars Dance

Venus and Mars form a conjunction at approximately two-year intervals—an average of 696 days. There were 212 conjunctions between 1900 and 2100. During the same time, there were only ninety-four oppositions—with an average time lapse of 770 days. Mars is retrograde 85 percent of the time when opposite Venus.[9] It is intriguing that in the twentieth century, there were only two oppositions with Venus in Gemini, Leo, Virgo, and Libra.

The greatest number was seven in Virgo. In the twenty-first century, there is only one opposition when Venus is in Taurus and Capricorn, but eight when she is in Aries.

Venus-Jupiter

Any aspect between the two benefic planets can be great. We expect more, are more optimistic, and tend to get what we expect in life. Jupiter is also linked to excesses, and when coupled with Venus we are talking food, wine, dance, and women or men. Too much of a good thing is not such a good thing.

The conjunctions and oppositions take place at thirteen-month intervals. Jupiter connections are the most predictable. However, some years we have only one, and other years there are up to three. Jupiter is always retrograde at the opposition, and Venus is retrograde at the conjunction at approximately forty-five-year intervals.

Venus-Saturn

We know that Saturn is about limitations, obstacles, and hindrances. Venus is about beauty, and Saturn really is not. With this connection, we tend to feel unattractive, undesirable, and not a whole lot of fun—old when we are young and getting more youthful with age. With age, we learn to accept that relationships require work, constant cultivation, nurturing, and learning to hug and show affection. Those with a Venus-Saturn connection can have successful relationships if they learn to accept themselves and others, flaws and all. They need to learn to accept that people like them just the way they are and not because they can do things for others. For some, there may be a string of relationships until the lessons of Saturn ("It takes time") take hold.

Saturn represents societal structures, thus Venus-Saturn individuals tend to think in terms of marriage, not an affair. If you are a woman, this may indicate that you are highly capable of looking after yourself, and therefore you want a partner with whom to share your life rather than someone who is not prepared to treat you as an equal. If you are a man, this aspect represents the women you are attracted to, and until you can accept the maturity, responsibility, and commitment given, you may find it difficult to stay in the relationship.

A Venus-Saturn connection is wonderful for manifesting what you desire, but it is not good for living from paycheck to paycheck. People with this aspect need to have rainy-day funds stashed in the mattress. That conviction that there is not enough is a wonderful motivator to help us retain rather than splurge.

This is a productive worker, ambitious about one area of life by location and several more through rulership. With the harsh aspects, it may take a long time to accept your lot in life, though with the soft ones, it is a given. This quote seems to sum it up: "The secret of contentment is knowing how to enjoy what you have, and to be able to lose all desire for things beyond your reach."[10] Learn to count your blessings. As a reminder, for those of us in love with someone with Venus-Saturn, they do not like to have their integrity questioned.

The Venus-Saturn Dance

We typically have to wait about 390 days for a Venus-Saturn conjunction. When Venus is retrograde during that time, there are three in a row. This only happens six times between 1900–2100: June 1916, February 1934, December 1954, March 1969, May 2060, and October 2098. There are two other times when there are three conjunctions in a row in the twenty-first century; however, Venus and Saturn are both direct. We would expect that the individuals who have a Venus-Saturn conjunction and were born during those unique times listed would have a different response to the energy. Saturn is always direct in motion at the conjunction. Some decades have more of these conjunctions—there were thirteen in the 1930s and only nine in the 1980s.

Venus-Uranus

With this aspect, we feel unique and individualistic, and we love our freedom to do as we please. We may delay or not want a committed relationship. The loss of independence is a frightening thought. When this duo is linked to the career and work sector, we are more willing to take risks, set up our own business, or take on jobs with a fluctuating income. We tend to be creative, innovative, and unpredictable. Our values, priorities, and personal assessment of the self are in for a radical overhaul in this life. If the aspect is harmonious, we accept these facets of our personalities with ease, whereas with the harsh aspects, we may think we need to fight for our right to be different.

In transit, Venus-Uranus in the fifth and eighth houses seems to be linked to fertility. The times when the New Venus period and the associated degrees activate our Venus-Uranus are times when we review our changes and opt to take more risks in order to reach what our heart desires.

Venus-Neptune

Neptune is linked to illusions, dreams, spirituality, and idealism. It can be deceptive, isolating, and disconnected from reality. We talk about seeing the world through rose-colored glasses. We all tend to fall in love with our senses, not our minds—marriage and love are two separate issues. Neptune is linked to hallucinations, drugs, and bliss.

When Venus is with Neptune, we may be in love with love. We may make the person fit our image rather than accept or see them as they are. Neptune sensitizes us, so we are more likely to have allergies to food or medication and to be more prone to want to escape reality. Our moods are also more likely to have the ebb and flow of the ocean; we sense the feelings of others—sometimes even correctly. It is easier to accept these qualities in ourselves if the aspects are soft sextiles, trines, quintiles, or deciles. With the squares and oppositions, we are more likely to resist—the stronger planet wins. If Neptune is the winner, escapism may become the solution. If Venus holds the stronger position, we may readily tap into our creativity, our loving and giving nature, and all of the Venusian qualities.

What could it mean if Venus stations on or makes conjunctions to your Neptune? This could awaken dormant talents linked to the planet Neptune. You could find yourself questioning your perception of reality, fall in love with love, or want to escape. Neptune is the great dissolver.

Venus-Pluto

This is an intensifying connection, with a strong potential for jealousy and possessiveness. Matters of the heart are taken seriously, and commitment is demanded—to one's own body and soul. Passion runs deep, and ownership is important. This connection can bring wealth and take it away. Edith Hathaway, a well-known Vedic astrologer, pointed out that in that system, Venus rules the eyes. We associate that x-ray[11] vision with Pluto; however, it is interesting to consider the phrase "if looks could kill" as a Venus-Pluto aspect. We astrologers gaze at the world differently. One night, a colleague and I watched someone walk in the dark wearing sunglasses and both commented that he had to be a Scorpio. Those smoldering eyes just have to belong to Venus-Pluto, from fire and ice to burning coals.

In the natural wheel using modern rulerships, these two planets oppose one another. Pluto is not known for his compromising demeanor, and we can have a stalemate. This could be a fantastic aspect for acquisition purposes, but a tad more challenging in relationships. Donald Trump has Venus in Cancer parallel Pluto in Leo and conjunct Saturn in Cancer; the phrase he wanted to trademark was "You're fired!" His TV show became extremely popular when the Venus occultation of 2004 took place on his Uranus at 17°53' Gemini; Uranus is conjunct the Sun in his chart.

1. The wheels in figures 25 and 26 show Scorpio at the nine o'clock, or Ascendant, position; the modern version places Aries here. For the purposes of these illustrations, this little variance makes no significant difference.
2. Maurice McCann, *The Sun & the Aspects* (London: Tara Astrological Publications, 2002), p. 34.
3. The declination varies a little between 23° N 26' to 23° N 27'; 23° N 28' is considered out of bounds of the Sun.
4. Sundials tell us what time it is as well as what season it is.
5. A great semicircle, which passes through a given point and the North and South celestial poles. Hour circles are perpendicular to the celestial equator.
6. Venus will be in Aries, Taurus, Gemini, Cancer, Leo, or Virgo.
7. Classical astrology also looks at mutual reception through exaltation, face, and term.
8. Table calculated with Kepler 7.0 program by Cosmic Patterns, Inc.
9. Venus was not retrograde at the opposition in this period.
10. Lin Yutang.
11. Uranus rules x-rays.

CHAPTER EIGHT

Charming Costumes

The distinct persona of Venus is amplified or muted by the zodiac outfit she is wearing. Venus in Aries holds some of the characteristics of Venus conjunct Mars and Venus in the first house. Venus in Gemini or Virgo has some of the traits of Venus conjunct Mercury or Venus in the third house. The actual planetary connections are stronger than the sign alone. Naturally, we also have the houses to add to the mix. Astrology is a wonderful language where we learn to superimpose qualities on the base or basics by adding layers and layers until we have the right words and qualities.

We all know something about the zodiac signs. We observe people with their Sun or another planet in a specific sign, and form opinions about the nature of the sign. In this chapter, we will hear Venus talk about what she desires by sign, and in first person, as a reminder that she can be self-absorbed. In terms of the cycle activating the sign, this can be the message: Do you desire these things now?

Venus in Earth: Taurus, Virgo, and Capricorn

Order is a lovely nymph, the child of Beauty and Wisdom; her attendants are
Comfort, Neatness, and Activity; her abode is the valley of happiness:
she is always to be found when sought for, and never appears as lovely as when
contrasted with her opponent, Disorder.
—Samuel Johnson

I take immense pleasure in the sensual side of life. I love to breathe in the scents of the flowers, the earth, and the distinct aroma of rain. A scent invokes powerful memories. I love the sense of touch, luxurious fabric, ripe fruit, and running my fingers on your skin. I'm here to learn to be grounded, reliable, and steady, and to sustain myself. I find comfort in stability and the knowledge that I have food and shelter and can indulge in simple pleasures. I strive for abundance, don't like waste, and have a knack for coping well in the material world. I love the taste of food—bitter, sour, sweet—all of it is the nectar of mother earth. I prefer to eat with my fingers, because that way I can also feel the food.

On the silver screen, I might be depicted wearing spectacles and a business suit (hopefully tweed or virgin wool) with my hair pulled into a severe bun. The minute the glasses are off, the hair is down, and the sexy top is on display, my sensual, sexy nature is revealed.

Venus in Taurus

I yearn for a piece of land, a garden for flowers, and a patch of dirt to grow my own food. It would be perfect with a picket fence around it. I rejoice in the bounty of the land. Food feeds all of our senses; beer and wine are libations of Venus. Moderation is difficult for me. I prefer to be dressed in rich velvet, plush satin, or brushed silk—it feels good on my skin and is comforting to the touch. I adore flowers; a potted rose lets me enjoy her beauty longer, and she will come back next season. I give them tender, loving care; it satisfies me to keep those flowers blooming longer.

I equate money, stocks and bonds, property, and a well-stocked pantry with security—I want it all and hold onto what I have. I don't mind if it takes time, but I don't like to settle or compromise my standards. My strong desire to have all of my needs met overrides expediency. I get pleasure out of giving gifts, preparing meals, and keeping my home comfortable and comforting. I do need to hear that my efforts are appreciated—a simple thank-you is all I want. I hate to be taken for granted; I may walk away.

I appreciate art in vibrant colors, a rich, textured oil painting rather than a light watercolor. I love to bring nature into my home, in natural tones of wooden cabinets, parquet floors, and surfaces pleasing to the touch. My home is welcoming, not sterile. A fireplace where I can burn wood rather than flick a switch to turn the gas on is more to my liking. I serve my meals on dishes that are serviceable and look good. The food needs to taste good, have substance, and be presented nicely—in that order. The fancy presentation of nouveau cuisine, where a few morsels of food look like a flower, does nothing for me—I prefer a hearty meal.

Venus in Taurus names: Catherine de Medici, Steffi Graf, Anne Frank, Anne Murray, Jessica Lange, Leonardo da Vinci, Billy Crystal, Adolf Hitler, Billy Joel, and Ryan O'Neal.

Venus in Virgo

I love to organize, perfect, and make things better. I'm logical and don't like to make choices based only on what my heart or instinct desires; I feel better if I analyze it myself first. I'm health-conscious; why would I feed myself "poison," which contaminates my body—my temple? I'm going to check the ingredients and those expiration dates. I love clothes that go anywhere and are washable; why add chemicals, when the fabric is against my sensitive skin? I have refined tastes and prefer classic clothing made of durable, natural fiber; everything matches neutral colors. Jewelry is a weakness of mine, but it has to be appraised and insured. I take care of my possessions, which incidentally don't own me.

I'm discerning, so only the tried, tested, and true meet my high standards. I'd rather do without than settle for second best. Of course, sometimes it is just as easy to improve and perfect that which has potential. I'm often depicted as a woman holding a sheaf of wheat, a caduceus, an autumn harvest, or a mortar and a pestle. I love to gain knowledge on nutrition and medicine. It is natural for me to be of service to others.

I like my home neat and organized, with everything in its place. In my kitchen, I want the clutter off the counters, so I can prepare my food on a clean surface. My kitchen is functional, with the best appliances, and any food I prepare is made with proper ingredients. I'd rather can my own vegetables and jams than chance adding chemicals to my food. I know that these are not good for me. I balance my checking account, and when it gets too low, I'm more frugal.

I hate to be criticized—keep it constructive. I get a little touchy if my choice of words or my understanding of them is undermined. I do have a tendency to correct the choice of your words. I don't like it if I'm made to feel unintelligent.

Venus in Virgo names: Ingrid Bergman, Rose Kennedy, Lucille Ball, Melanie Griffith, Roger Moore, Henry Ford, Sean Lennon, and Luciano Pavarotti.

Venus in Capricorn

I'm a lady, with the classic, little black designer dress and a business suit—indeed, I have the right attire for any occasion. I'm not all business, but I will not reveal my sensual side to just anyone. I'm simply too sensible for that and I do need to get to know you first. I might be called the original material girl. Incidentally, Madonna, who sang that song, has Venus in Leo.[1] I grew up early and had to assume the role of the parent or the responsible one in childhood.

I'm ambitious about matters connected to the astrological house I live in and my connections (aspects). I excel at making those connections; I have a flair for what is essential. I'm capable of managing anything I choose, to perfection. I know how to set attainable goals and do the work, plus I function well in the real world. In fact, if there is no work involved, it may seem unearned—I prefer to earn my rewards. My sensitivities are mine to control. It takes the same kind of executive ability to run a business or a household; I can do both. I hate it if my integrity is questioned. I'm honest to the core; please respect my boundaries just as I respect yours.

My house needs to accommodate my business activities. It needs to be functional, with space for an office, as I may bring work home or work from its comfort. I want to know where everything is and who is taking care of what. Yes, I delegate, as my time is better spent doing something where my skills are the best. When I cook, it is excellent, but I may opt to have someone else do that task. I earn the money to pay for simple luxuries. I don't compromise on skin care products; skin is our biggest organ, and we only get one. I may not be perfect, but I'm working on it. In time, I'll learn to accept that others might love me for who I am rather than what I can do for them.

Venus in Capricorn names: Louisa May Alcott, Britney Spears, Tyra Banks, Caroline Kennedy, Ronald Reagan, Paul Newman, George Patton, and Frank Sinatra.

Venus Cycle in Earth

Nature does nothing uselessly.
—Aristotle

The Venus retrograde and direct stations take place on either side of the New Venus. Out of the five cycles, only one remains completely in earth during the first four decades of the twenty-first century. That is the one in Capricorn. (There are two retrograde stations in Virgo, the final one in 2015; however, Venus spends the bulk of that time in Leo.) Venus direct stations in Taurus begin in 2044. The first New Venus in Taurus is in 2076—the Virgo one begins anew in the twenty-second century.

What could we expect in our lives when Venus spends four months in Capricorn marking three degrees? The house placement and aspects to natal or progressed planets are typically more crucial. The sign placement adds its mode of operation to the mix. The element of earth has a message: "Roll up your sleeves and do the work!" Capricorn is the ultimate executive of the zodiac. There is a business-and-action plan starting with a mission statement. Every eight years there is a repeat performance. Every four years following the New Venus there is a Full Venus within a degree of it to entice us some more.

Capricorn sticks to facts, and talks about what is rather than what could be. It is about status and career, plus obstacles on our path and the hard work needed to overcome these. We learn about patience and prudence—we can proceed only when we have all the facts. This can be immobilizing, but also an antidote to laziness, inertia, and a lack of direction. At first, we may feel overwhelmed, but need go back to learn to do things properly. Maybe we just need to learn to appreciate our innate talents and ability to handle the administration of finances (second and eighth houses) or daily routines (third and sixth houses). We ask ourselves, is it that we are what we do or that we do what we are—and what are we?

Taurus might pose questions about its true value—is it worth the time and effort? Virgo might want to know if it is perfect enough and how to make it better.

Venus in Air: Gemini, Libra, and Aquarius

One must have sunshine, freedom and a little flower.
—Hans Christian Andersen

I love thoughts, ideas, conversations, and good company with interesting people. I love to hear and read sweet nothings, but I also need deep and profound ideas to ponder. I may appear easy, breezy, and even flirtatious, but don't mistake that for shallowness. I'm inquisitive and genuinely interested in what you think. I need to connect with others on a mental level, so I can learn to understand how I feel and what I truly value. That quote that men are turned on by what they see, and women by what they hear, might have been coined by someone with Venus in an air sign. I actually tend to fall in love with the mind more than with the body.

I love flowing repartee, and even clothes should not hinder movement. Give me loose, flowing garments in the latest fashion colors; what I wore last season no longer pleases my senses. I adore fresh experiences and want to keep all my relationships that way. I'm ever changing because I want to keep others interested—I'd hate it if you became bored with me, or me with you. I'm a free spirit; perhaps I was the star in *Gone with the Wind*.

Venus in Libra

I'm charming, nice, and at ease. I'm confident in who I am, and have an innate knack for getting cooperation. Yes, I can be very persuasive. I don't like to offend others; first, it is in bad taste, and second, you do catch more bees with honey than with vinegar. I'm happy to discuss it amicably and reach a mutually acceptable solution. I have earned the monikers of indecisive and lazy. Let me explain. To be perfectly just and fair takes time. I can see both sides, and mostly each view has equal merit. Choosing between two things of equal value is tough for most of us, so I find it a little harder, but doable. I don't like to waste energy. I plan, then execute; there is absolutely no need to break into a sweat over anything. Women don't sweat, they merely mist up; so say the Southern belles. What you call laziness, I call economy of action. I'm the one behind the politically correct words—we should not be judgmental. Granted, the world may have gone a little overboard with some of them, which was not my intent. I'm never brazen or brash.

I abhor arguments and fights, but I'm extremely skilled at debate and excel at starting pro and con arguments. I love banter and like to play the mediator and to watch how it plays out, with nothing invested in the outcome. What is just and fair is the role of Lady Justice. In reality, justice may be in principle only, but I'm the best equipped to debate that question.

I love a sense of space and beautiful pictures that are pleasing to the eye. A white wall is so soothing, and just a dab of color brings that to life. Clashing colors upset my ethereal sense of beauty. I want a floor plan where the chi flows; give me wind chimes, water fountains, and beautifully arranged bouquets of flowers.

Venus in Libra names: Connie Chung, Jenna Elfman, Calista Flockhart, Georgia O'Keefe, Sean Connery, Barry Gibb, Samuel Goldwyn, and Oscar Wilde.

Venus in Gemini

I love conversations and collecting news, information, and new knowledge—I share it uncensored. Teach me something I don't know. I have a reputation for being flirtatious, curious, inquisitive, chatty, cheerful, witty, and vivacious. I love new experiences, new learning, and keeping my mind busy with thoughts, books, movies, crossword puzzles, and even trivia. Every day is a new lifetime! I also love camaraderie in all my relationships. Perhaps I'm like the butterfly, flitting from flower to flower to collect nectar. Did you know that a butterfly can perceive ultraviolet wavelengths of light? I want to learn to tune in to the higher octaves in my quest for wisdom. Life is short; I want to inhale every moment. I want to hear noise, music, silence, roaring laughter, thunder, and lightning as well as gentle rain.

My home needs to have light and air, plenty of windows, and room for me to dance and move about unhindered. I hate to be crowded, although I don't mind being in a crowd, as it is stimulating. I don't have to own my own home nor do I yearn for a plot of land—city dwelling keeps me close to all the action. I'm often out and about anyway, and what if I want to move? Giving notice is so much more expedient than waiting to sell a place.

I do like my gadgets, my cell phone, laptop, and remote controls for the TV and CD player and even one for the fan to keep the air flowing. I don't like to be plugged into the wall; I love to walk and talk. I love sending and getting little notes—the faster the better of course—but pretty cards through the mail with nice little sayings do give me pleasure. I can curl up with a book or a good show, but I also like to have several things going at the same time. I can talk and fill in a crossword puzzle at the same time. I'd hate to be thought of as uninteresting.

Venus in Gemini names: Bette Davis, Ann Landers, Eva Perón, Venus Williams, Johannes Brahms, Bob Dylan, Henry VIII, and Joe Torre.

Venus in Aquarius

I'm cerebral. I love sharing my wisdom and truth with the world. My personal freedom is important to me; I don't like being boxed in. I appreciate friendship and the lively exchange of ideas and concepts. I think I'm somewhat unusual and perhaps even eccentric, and I like similar qualities in others. I don't like it if things become stagnant; I prefer to discover new ways of looking at issues in life.

I adore intellectual stimulation, but I'm after the truth, not just information. It is that Utopian mind-meld as in *Star Trek* that I find intriguing and stimulating. My thirst for knowledge and truth is never quenched. I'll dig deep to find what motivates me and those I find intriguing. I'm not particularly jealous or possessive, and those traits disturb me in others.

I may seem aloof, but beneath the surface I'm emotional and need assurance that I'm interesting and unique. I tend to be more social in groups than in one-on-one encounters. There is safety in numbers, and I don't like to reveal my vulnerable side all that quickly. It is fairly easy to hold my interest; simply have personality traits and quirks that are different, be emotional or intense, or be the drama queen and you will be irresistible to the intrigued Aquarius. Try not to tell the same story too many times; I did listen the first time.

I take pride in my honesty—I hate to have that questioned. I tend to react with vehemence and sometimes even vengeance. I'm straightforward, and I don't handle the opposite in others very well.

Venus in Aquarius names: Natalie Cole, Ellen DeGeneres, Gloria Steinem, Kiri Te Kanawa, Nicolas Cage, Mel Gibson, Ricky Martin, Rudolf Steiner, and J.R.R. Tolkien.

Venus Cycle in Air

The only active cycle in the element of air takes place in Gemini. Libra enters the scene beginning with the direct station in 2010, Full Venus in 2022, and New Venus in 2034. Her cycles shift gently. The next Aquarius cycle also begins with a direct station in 2062, Full Venus in 2081, and New Venus in 2085. The potent Gemini cycle is connected to the North Node of Venus, the sign of her occultation, which brings about societal change. (See chapter 12.)

The important thing to note is that the essence of Gemini is to share knowledge and information without adding any analysis to it. The thirst for information is so deep in Gemini that it fails to recognize any language or communication boundaries.

Where is inquisitive Gemini in your chart? What does she want you to discuss and learn? Where is it that you need to be open-minded and not allow a language barrier to keep you from gaining more insights into your life? Where in your life could you be spreading the word? Take note of transiting New and Full Moons, planetary stations, and even transiting conjunctions to the Venus degrees. These tend to activate the earlier Venus degrees, which hold the note through ages.

Venus in Fire: Aries, Leo, and Sagittarius

I decided that it was not wisdom that enabled [poets] to write their poetry, but a kind of instinct or inspiration, such as you find in seers and prophets who deliver all their sublime messages without knowing in the least what they mean.
—Socrates

I must admit that I love drama. The world is a stage, and I have a role to play. I'm encouraging, inspired, and have a "can do" attitude. I have energy galore, and at times I'm like a puppy jumping on all fours, anxious to get moving. Inactivity and apathy are feelings I don't have time to experience. I hate to plod along; I'd rather create a scene, start a new activity, or laugh aloud—it's better than crying.

Ready, set, go, or lights, camera, action! I love performing. Life is a play, and I got the lead role. I tend to be a little feisty—maybe I had the lead role in *Who Is Afraid of Virginia Woolf?* On one hand, I love those grand gestures, but on the other, flowers from a friend give me mixed messages. I secretly crave applause and approval. That's just between you and me, right?

Fire is the energy that fuels the Sun and ignites that flame of inspiration within us. It is creative energy, and it regenerates us. Think of forest fires—only a few seasons later, the underbrush is gone and there is fresh, new vegetation for wild animals to graze on, and there are new, regenerated forests springing up. Pine trees don't pollinate without the help of a fire; it is one of nature's ways to foster new growth. The same theme applies to fire in the astrological language—it is intended to inspire us to venture toward new things.

Venus in Aries

I'm impetuous and daring and tire of things easily. I adore spontaneity; I'm always up for trying something new. I like to launch new projects; they hold my interest for a while. I know how to find a person to take over and nurture these ideas to grow. If I can find new ways to tackle the mundane, I don't mind it at all—the tedium of routines brings out the worst in me. I'm not all that selfish; I simply need to please myself first. Once I'm happy and in my element, I'm a lot of fun. I'm passionate and inspired and can spark a fire within others.

I do love adventure, bold colors, and startling contrasts, and I dress to please myself. I'm drawn to relationships; I don't like to be alone, but I'm not in the least clingy. I have to sow my wild oats before settling down. I like to try new things; sitting still is not my strong suit—I may want to climb that mountain, try parachuting or deep sea diving, and explore other countries, lifestyles, and foods.

Money buys security and lets me have my toys. Home is where I hang my hat, but the setting can change a lot. I buy things on an impulse, when the object strikes my fancy. It costs what it costs, after the haggling, and I'll earn the funds when I can. I get upset quickly and forget the storm with equal ease. My desire is to leap into action and fix whatever upset those I love or me.

I love my car, sports equipment, and my sporty outfits. Detail my car, take me out to the ball game, cook me a meal, or surprise me with something new. Sitting still is not my strong suit; without action I wither away.

Venus in Aries names: Mario Andretti, Harry Belafonte, Shirley Temple-Black, Sigmund Freud, Betty Friedan, Jimmy Hoffa, Priscilla Presley, and Queen Victoria of England.

Venus in Leo

> *Think highly of yourself because the world takes you at your own estimate.*
> —Anonymous

I'm enthusiastic, boisterous, inspired, and gregarious. I adore attention. That primal fire of life force burns strong in my heart. I hold onto the heat of the Sun and do everything with passion, and I want to ignite the same in others. I warm up the room just by being in it. Yes, I have a temper—fire burns, warms, scorches, and rejuvenates. I like my fun and recreation; I want to play like children do—with abandon and joy. Lights, camera, action!

I enjoy being at center stage, but want to ensure that everyone is having fun. I'm a wonderful social director and don't like to have life stagnate to humdrum repetition. I'll chance it and take a few risks; you can't win if you don't play!

I want my home to be my castle. I love both comfort and luxury. Art, paintings, sculptures, and trinkets created by me and those I know brighten up my home. I'm actually a great big pussy cat; I purr but also snarl—try not to stroke my fur against the grain. The older I get, the harder it is to placate me with praise. Keep it sincere—I do. I love kids, sports, and fun. I'm creative and want to inspire creativity in others. I radiate warmth and lighten up any space I'm in. I like freedom but don't like to be alone; a small audience is always nice. I can be stubborn and inflexible. I know what I like and don't like.

Venus in Leo names: Phyllis Diller, Van Morrison, Olivia Newton-John, Peter Paul Rubens, Pete Sampras, Mika Waltari, and Barbara Walters.

Venus in Sagittarius

I'll have that big piece of cheesecake, please! I want to have a go first—at the bat, on the stage, or at any starting gate. I burn all I consume—calories are energy—and eating on the run is great for my digestion. A rolling stone gathers no moss, and I'm much the same. I love sports, fun, philosophy, and exploring the world, and it's a big, wonderful world out there. I love to laugh; I'll even tell myself jokes just to brighten up the day. I'm happy-go-lucky most of the time. I want to know what makes me tick, how the world works, and your opinion on any topic.

I love space around me; I want to be free to roam, play, and have loads of fun. I don't make little plans—mine are always grand. I might not realize them quickly, but I can dream, can't I? Nature would not entice us to have dreams if they couldn't be realized—hah, I'll dream even bigger dreams next time around! I can be goofy and spiritual. It is the intent that matters; I intend to be all I can be and then some.

I have my convictions and can be opinionated—I have an opinion on most everything. How can we understand anything if we don't ponder what the words, ideas, and notions truly mean? Possessions aren't that important—well, I need to have my toys: bats, racquets, books on wisdom, and a few collectibles from the world of myth. Doesn't everyone want to have a Yoda, Merlin, Mickey Mouse, angel, or fairy?

Venus in Leo names: Charles de Gaulle, Emily Dickinson, Jane Fonda, Joe DiMaggio, Charles A. Lindbergh, Charles M. Schultz, and Frances Yates.

Venus Cycle in Fire

Two out of the five Venus cycles are presently in fire—Aries until the 2030s and Leo to 2095. Sagittarius begins in 2064 with a direct station.

Each cycle begins with Evening Star Venus waning at the retrograde station. She is recharging for the impulse to initiate new creative endeavors after the direct station. She strives for recognition to build up confidence. The impetus of fire is to ignite that fire in our bellies and in our hearts. Aries is the independent one, wanting to be the first, and is not called the pioneer by chance. A divine spark burns within us all—Aries a flint that bursts into a flame, Leo the bonfire, and Sagittarius an eternal flame, which cannot be extinguished. Fire is creative work, excitement, exaggeration, passion, entertaining, and showing off. With this cycle, we are inspired to begin new projects, to find new, innovative ways to handle the issues by sign, house, and aspect.

The direct station of Venus in fire is good for launching creative projects. As with the New Moon, it takes time for the idea or notion to grow roots. However, four years later, the Full Venus will bring welcome praise.

Venus in Water: Cancer, Scorpio, and Pisces

A person has three choices in life. You can swim against the tide and get exhausted, or you can tread water and let the tide sweep you away, or you can swim with the tide, and let it take you where it wants you to go.
—Diane Frolov and Andrew Schneider

I'm private and guarded until I get to know you. I don't like strange settings, and it takes me time to familiarize myself with a new environment. Until I feel safe, I tend to appear shy and reserved. The atmosphere has its moods, and I sense these. I'm emotional, which is sometimes spelled moody. Lest we forget, emotions do not simply mean sensitive and sympathetic. My emotions range from love to hate, desire to repulsion, pleasure to pain—sometimes the time it takes for me to go from one to the other can be measured in nanoseconds. I'm nostalgic and have fears and deep memories. I have premonitions, which can feel as real as something that was committed onto paper. I'm persistent and powerful; just think how water in nature can erode the hardest rock. It simply takes a little longer—time is not linear!

I love and hate with equal passion. I can go from tenderness to harshness in the blink of an eye. I'm sentimental and sympathetic; I sense how others feel. I cry and laugh—at times I do not know why the tears flow or what brought on the laughter.

In a perfect world, my home is near water—a river, a lake, an ocean, or a bottomless forest pond. In a pinch, I'll settle for a fountain. My bathroom is my oasis, as water and tears are cleansing. I guard my home, its privacy, and my family; what is more important than the material contents is what is within these walls. Sentimental value is always more precious to me than the actual monetary price tags we place on our possessions.

Venus in Cancer

I love security, safety, and family. I need to have a home to call my own; I do prefer to have a title to it rather than know I might need to move. I'm sensitive, caring, and compassionate. I'll cry with you and on your behalf. I sense the feelings of others, both pain and pleasure. I adore sitting in water; it melts those worries away. I tend to fret and know what is to be long before the actual event. I'm tuned in to the thoughts, anxiety, and feelings of others—if I could fix it, I would.

Mementos have a place of honor in my home and heart—lace, satin, family photos, and images of those special moments that make life worthwhile. I need to retreat and be by myself to sort through those deep feelings. I'm thrifty—waste not, want not is my motto. I keep insurance against those rainy days; it may be a spare set of china or cash in a safe place in addition to the bank. I've been known to stash money, bonds, certificates, and policies in the sock drawer or with the oven trays. My pantry is well stocked; we need food to survive. I love for keeps.

I'll nurture and nourish those I consider my family. I'm sentimental and love to reminisce. There is something comforting about traditions and the support of family.

I love to surround myself with heirlooms; grandma's jewels and grandpa's tools are tended to with loving care. I love to nest. Even when I am the head of a huge company, I consider my employees my family and will remember their birthdays and special occasions. It's natural for me to make sure everyone is comfortable and not ill at ease.

Venus in Cancer names: Halle Berry, Cameron Diaz, Richard Cheney, Bob Hope, Tom Jones, Vincent Price, Nelson Rockefeller, and Maureen Stapleton.

Venus in Scorpio

I'm deep, intense, and passionate, just in case you didn't notice. I can be both obsessive and possessive; I bond with others through time. I hate shallow camaraderie, superficial conversation, and flippant shows of affection. I'm not that easy to get to know. I'm wary, guarded, and restrained until I know I can trust you. I'll trust and respect you once you have earned it. I'm extremely protective of those I love; you can call me a name and get away with it, but do not think you can talk about those I love in that sort of fashion.

I love with passion and hate with vengeance—the line between any two extremes is mighty thin. I can turn to ice and surgically remove you from my life—you simply cease to exist in my reality. I do not do things on a whim. If I feel hurt, I brood and feel wounded; however, I won't share that with you. I lock my secrets and those of others in a vault. In my quest to understand what motivates me, I scrutinize the motivation of others. It may seem as though I look right through you, but having been to the depths of desperation and the heights of ecstasy, I have concluded that most things in life are a matter of life and death.

I like my possessions, always have money tucked aside, and am generous with my resources when we share a familial bond. Give me a secluded home, guarded by the deep forest or a wonderful tall hedge, and I feel safe. I do love my alone time more than most. I'm actually not particularly social, but one-on-one I give others my undivided attention.

Venus in Scorpio names: Stephen Arroyo, Dr. Joyce Brothers, Sylvia Browne, Hillary Clinton, Leonardo DiCaprio, Mohandas Gandhi, Paul Getty, Thor Heyerdahl, Tatum O'Neal, plus numerous astrologers and notorious individuals.

Venus in Pisces

My emotions and feelings bubble up from the sea of humanity—from the very depths of the ocean. I'm sensitive, caring, kind, and vulnerable. I'm fluid, absorbing, and malleable, readily adapting to my environment. I take in the energy of others, but also the energy of the planets that surround me, or the house I reside in. I can be what you would like me to be—it's my shield against hurt and pain. I sense the emotions and moods of those around me. I can hide my vulnerability and hurts, because I don't want to expose myself to insensitivity. Good deeds and kindness toward humanity have always been part of my inner world—random acts of kindness are something I adore. I need to understand my deep feelings and faith in the unseen.

I may be dreamy, idealistic, and passionate about my beliefs. I don't care what you had for lunch; I want to know how you feel about world hunger or injustice. I remember my first haircut—it hurt, and everyone laughed at my tears. That was the last time I openly showed my feelings. Why is Venus exalted in Pisces? I'll tell you. I love others more than I love myself. I care for humanity and would give others the shirt off my back—not out of generosity, but because someone else needs it more than I do.

When I am with others, part of them becomes a part of me and a part of me stays with them. It's another safeguard so I won't need to feel my passionate commitment all at once. Love is a narcotic, and narcotics are a way to escape my profound sensitivities for a brief while. I weep for humanity as passionately as I weep for myself. I'm at my strongest when I am protecting another vulnerable soul.

Venus in Pisces names: Ursula Andress, Johann Sebastian Bach, Edgar Cayce, Charles Dickens, Ali MacGraw, Michelle Pfeiffer, Barbra Streisand, Julia Louis-Dreyfus.

Venus Cycle in Water

The Scorpio cycle ends with the retrograde station in 2050. The Cancer cycle gave way to the current Gemini cycle; there is a new one beginning in 2071. The Pisces one concluded in 1910 with a retrograde station; a new cycle in Pisces begins with a direct station in 2009, and the first Full Venus in a late degree of Pisces takes place in 2045.

Water is a forceful element that flows around obstacles. It wears away solid structures, extinguishes fire, and fills the air with moisture. Our bodies are 95 percent water; we feel the change of tides. Do you feel in tune with your desires, possessions, and values? What needs to be worn away? Do you yearn for a deeper connection? Scorpio is about elimination, possessions, shared resources, and commitment. Are you prepared to commit to the issues by house and aspect?

All of the retrograde periods are review times in our lives. The Venus question "Does something still hold value to you?" could be about your closest relationship or that brand-new vehicle in the driveway. If what you have is deemed unimportant by you but not others, you will eliminate it. Other people, by their actions and words, will influence your thought processes, but ultimately the decisions are yours based on your own values.

With Libra, we meet others, and with Scorpio, we commit to them. To go against the flow now would be to start a new relationship for all the wrong reasons or to end one for a different set of wrong reasons. This cycle offers us the tools to add strength to our relationships by eliminating behaviors and attitudes that no longer serve a valid purpose. Hasty decisions with financial matters and taxes are not advisable either. Wherever possible, defer decisions until you have had plenty of time to consider your choices. Scorpio is the symbolic representation of death, which represents where we are letting go of things and issues that are no longer feasible, those that cannot survive. On the spiritual side, this is a wonderful period for reassessing your values, needs, and desires by carrying out a cyclical clean-up. It is excellent for reassessing what you long for in a specific area of your life.

Quick Review of Venus in the Signs

Material Venus in an earth sign, in a house of substance (houses two, six, and ten), and in connection with Saturn is blessed (or cursed) with common sense and therefore less likely to act on a whim or throw caution to the wind. Building anything concrete or a legacy to leave behind would be a desired achievement for the pragmatic, traditional, and authoritative earth Venus.

Venus in a fire sign, in a house of life (houses one, five, and nine), or conjunct the Sun, Mars, or Jupiter is blessed with enthusiasm and inspiration, that never-give-up approach to life. Caution is considered a curse word—to live is to explore and expand one's horizons! Having a vision and sparking that proverbial fire in the belly in all of those around this feisty Venus type would feed her need for recognition.

Venus in a water sign, in a house of endings (houses four, eight, and twelve), or in connection with the Moon, Neptune, or Pluto is blessed with a high emotional quotient and the ability to understand others instinctively. To her, fulfillment is achieved by bonding with others and learning to understand herself better. She is here to build memories, to rely on feelings more than logic, and to learn to trust her own intuition.

Intellectual Venus in an air sign, in a house of relationships (houses three, seven, and eleven), or in contact with Mercury or Uranus is blessed with imaginative thoughts and an inventive mindset. If something doesn't work, she will try a new approach. She looks for solutions through her interactions with others. She thrives on the energy of others, particularly when new knowledge and information are to be gained.

Every trait, characteristic, and quality has its polar opposite. We recognize the traits in others that also reside within us. Here is a little story to illustrate this point.

Two Wolves

One evening, an old Cherokee told his grandson about a debate that goes on inside people. He said, "My son, the battle is between two 'wolves' inside us all. One is evil. It is anger, envy, jealousy, sorrow, regret, greed, arrogance, self-pity, guilt, resentment, inferiority, lies, false pride, superiority, and ego. The other is good. It is joy, peace, love, hope, serenity, humility, kindness, benevolence, empathy, generosity, truth, compassion, and faith." The grandson thought about it for a minute and then asked his grandfather, "Which wolf wins?" The old Cherokee simply replied, "The one you feed."

---

1. Born August 16, 1958, at 7:05 a.m.—DD. Conflicting records show 1959; however, her sister Paula was born that year. In addition, Madonna states that she was the oldest of the girls. Madonna also stated that she has Aquarius rising (source: AstroDatabank, Erin Sullivan quotes Helen Stillwell), which would actually mean she was born at 19:05 (7:05 p.m.).

CHAPTER NINE

Venus Retrograde: The Magnifying Glass

Mirror, mirror on the wall, who is the fairest of them all?

The above query is a line from the fairy tale *Snow White* uttered by the jealous stepmother. Venus is linked to beauty in every shape and form. Women are supposed to be beautiful. When we pick up any magazine or turn on the TV, there it is glaring at us—the message that we need to be young, flawless, and ever beautiful. Remember that the retrograde Venus station started the process that resulted in the book you are reading. Over the past sixteen years—two Venus cycles—many a client with Venus retrograde has entered my life. The first lesson this brought was to have a box of tissues handy. Whenever a person born with Venus retrograde encounters another human being who seems to understand, the tears begin to flow.

The magnifying surface of a barber's mirror is a wonderful image for illustrating the Venus retrograde phenomenon. We view life blown out of proportion; issues that may be irrelevant to others gain significance. We tend to recoil when gazing into the magnified image of our face until we learn to accept ourselves, flaws and all. Until then, we stay busy trying to fix or rectify any perceived flaw.

If you were born with Venus retrograde, there is often the issue of not feeling loved by Mother or not loving her. Other issues include not feeling appreciated by your partner or worthy of a decent job. When Venus is retrograde, she is closest to the Earth. When a planet is closer in proximity, the issues and things it represents take on stronger meaning. The effects are more pronounced. All Venus things are up close and intimately personal. Your feelings tend to be intense, and you may feel that you are not understood and that your values, needs, and priorities are not as important as those of others. The longer we think or feel a certain way, the more it becomes the reality.

If thou findest it not within, thou will never find it without.
—Anonymous

People born with Venus retrograde may feel they aren't perfect enough, not lovable enough, not accommodating enough, and not nice enough nor worthy of things others take for granted. No amount of external reassurance will make individuals with Venus in this condition appreciate or value a quality they feels isn't theirs. The sense of validation needs to come from within.

Life does not change us, it unfolds us.
—Max Fleisch

The symbol of Venus cycles in the sky is the rose. A rose bud is associated with budding love. Slowly the petals unfold to reveal the beauty of the flower within. The rose is probably the most cultivated flower. It comes in a variety of colors. Recently, the local florist shop near my home started carrying variegated versions. There was one in deep reds with pinkish lines called Intuition—a very regal looking flower, whose essence was impossible to capture on film. Roses, just like people, come in a variety of shades, colors, and sizes, from wild roses to ones designed for specific occasions. Likewise, the qualities of the Venus in each chart are unique, each one of us being equally unique.

Venus represents our inherent values and desires. The cycle of Venus unfolds slowly. Those of us born with Venus retrograde have a more permanent or undeviating value system with which we need to bond.

When we talk about retrograde periods, we use words such as review, revise, reevaluate, rekindle, repay, repair, and reassess. These are Venus-type words, rather than the words repair, redo, and rethink, which resonate with Mercury. How long you have to work on reassessing your personal values and priorities depends on the condition of your Venus. If you were born with Venus retrograde, you have a lifetime to review and revise your personal evaluation of your worth. If Venus turns retrograde by progression in your chart, you have forty years of inner searching of your values. The shortest period is the forty days when Venus is retrograde in transit. This can seem like an eternity when we grabble with issues that only we can resolve. Eventually, via progression, Venus turns direct in a natal chart by the mid-forties at the latest, reflecting a significant shift in life.

Venus also represents social skills and a sense of belonging. A person with Venus retrograde may feel a strong sense of disconnectedness—not belonging anywhere.

We all talk about others being a reflection of us. We do not recognize traits we do not have, but only those we own—good, bad, or indifferent. Think back to when you were first in a couple with someone, and how you stood in front of the mirror thinking you made a perfect couple. Venus is about love and relationships, and during the retrograde cycle, it is as if we stare into that magnifying mirror to notice any flaws, real or blown out of proportion. If everything is in accordance with our values and priorities, we note nothing wrong, but if not, then we need to take a good, honest look at what is happening in our relationship.

Inner refuge is refuge in ourselves, in our ultimate potential. When we recognize and nourish this potential, we have found the real meaning of refuge.
—Kathleen McDonald

Through the retrograde phenomenon, we learn to look at and scrutinize all of these issues. The flaws and shortcomings are in full view. It is no longer about what is on the surface, but the motivation behind it all. Are we feeling loved and appreciated? Are our needs and desires being met? Are we receiving what we think and feel we are worth? Are we worthy of more? Does the love of our life meet or exceed our expectations? Does the object of our adoration have clay feet? Do we get shows of affection or rejection?

We typically acknowledge that the outer planets, which move along very slowly, describe processes. The outer planets in transit (and natally for that matter), which stay in the same place for a long time, are similar to house guests who are beginning to wear out their welcome. We have to share a space with them that would normally be all our own. For example, no matter how much you love your sister, when she drops in unannounced for a short visit and is still there after three months, you could probably strangle her.

The Dance of the Seven Veils

In mythology, the Venus archetype is stripped of her valued possessions. She is forced to enter the underworld vulnerable and exposed, without clothing or regalia to indicate her worldly status. So, who are you when no one is watching? Can you love yourself without external validation? A natal Venus retrograde is either an Evening Star or a Morning Star. The nature of each is different. The Evening Star is no longer illuminated by the light of the Sun, the ego. The sense of worth is fragile and easily bruised. The Evening Star phase seems to be more prominently linked to relationships; the Libra and seventh-house connections are strong. We strive harder to please others. The Morning Star Venus appears to be connected more strongly to Taurus and the second house. There can be an immense inner motivation to prove our worth through acquisitions or to completely denounce the importance of material wealth.

With the retrograde periods, we talk about the journey into the underworld, which we understand to be our subconscious or unconscious mind. Once there, we encounter some hidden or forgotten truths about ourselves and ideally emerge stronger and better. When we are talking about Venus, we are invoking something in our hearts. Traditionally, we link the heart to the Sun and Leo, which is the pump (or engine) that physically fuels all our organs. However, we link the heart chakra to Venus and assign it the Venus color of pink. Venus is about what the heart desires, and in matters of the heart, Venus typically wins.

We typically associate retrograde planets with karma and may believe that we failed in some regard in a previous lifetime. An old client told the most heart-wrenching story about her childhood (Venus retrograde in Scorpio in the first house rising before the Sun). She recalled the many social events her charming, beautiful mother hosted. The mother would play the piano to a gathering in the affluent surroundings. There were also servants in the picture. She was not allowed to be around the beautiful people; rather, she was "imprisoned" in her room. There was to be no leaving her bed to use the bathroom or get a snack. On occasion, she recalls being strapped onto her bed. The mother thought that her clumsy, unattractive daughter was not good enough to introduce to her circle of friends. She felt that her mother absolutely hated her.

Later, after having moved away from her native land, she married a man who had numerous affairs, and she was always made to feel ugly, unintelligent, and not in the least feminine. In her senior years, she divorced her husband, took up photography, and finds solace in nature. Her main comment about her Venus is that she truly feels she does not belong anywhere and no one truly appreciates her.

Many of the Venus retrograde individuals come from an affluent background—homes with servants to help out, the best schools, private tutors—in short, a privileged childhood. Similarly to the previous story, many a Venus retrograde individual talks about being raised to be better than the rest. Often, these individuals are rather gifted in the arts; they got the best teachers. Some are exceptionally intuitive and sensitive. They feel and sense things differently. Perhaps they have synesthesia, where one or more senses blur together. That kind of sensitivity can make us allergic to perfumes or even the odor of a person.

As Venus is linked to marriage and pregnancy, there is a strong probability of having more than one marriage, and the risk of miscarriage is higher. There can be reversals in finances; many of these individuals are children of wealthy and well-heeled parents. Money is not everything to them; rather, happiness and contentment are more important.

Here are some famous Venus retrograde names with their Venus sign placement: Ansel Adams (photographer/artist, Aquarius), Kathy Bates (actress, Cancer), Milton Berle (comedian, Cancer), Annie Besant (social reformer and feminist, Libra), Lizzie Borden (acquitted in the homicide of her parents, Cancer), Elizabeth Barrett Browning (writer, Pisces), Charlie Chaplin (comedian, Taurus), Morgan Fairchild (actress, Aquarius), Jodie Foster (actress and director, Scorpio), David Koresh (cult leader, Virgo), Graham Nash (musician, Aquarius), Chris Rock (comedian, Capricorn), Donald Rumsfeld (government official, Cancer), Ted Turner (media mogul, Scorpio), and Peter Ustinov (actor, Taurus).

Elizabeth Barrett Browning, the author, was also from a well-to-do family. Her father would not allow her or her sisters to marry, so she eloped at age forty and finally found happiness. She was frail—her Venus an Evening Star—and had several miscarriages before giving birth to one daughter. Her Venus stationed direct when she was twenty-nine, and her progressed Venus was conjunct her natal Sun when she was forty. That progression is a typical indication of falling in love, which we will discuss in chapter 11. Venus seldom denies but rather delays, so that when you achieve your heart's desire, you appreciate it so much more.

Likability Factor

"Likeability is truly the secret of a charmed, happy and profitable life," writes Tim Sanders in his book *The Likeability Factor*.[1] It is all common sense, but in terms of the elements we use in astrology, the concepts he brings up are intriguing. Friendliness is the ability to communicate liking and openness to others—sounds like air. Relevance is how capable we are in connecting with the interests, needs, and wants of others—sounds like fire. Empathy is the ability to acknowledge the feelings of others—water. Realness is equated with accountability; we need to keep our promises and have integrity—earth.

Venus retrograde individuals may feel that they are short on likability and that they have to work harder to please. Deena Grier, a Montreal-based astrologer who passed away shortly after writing an article on Venus retrograde, which was published in the April/May 2003 issue of the *Guild News*, wrote: "In childhood these individuals are shy and socially inept, but oddly bright—picking up on things that will make adults remark with delight. It is not that the Venus retrograde individual does not know how to give affection—sometimes the person gives too much and it isn't received. Their taste in the opposite or same sex is always going to go against the norm. Men with this configuration often tune into women quite easily."[2]

Deena concludes that we need to listen more closely to these individuals, a statement with which I absolutely concur. Please do not tell them that they did not know how to love and share in a previous life—the problem could be rooted in childhood. Further wounding someone with such a sensitive and sensitized nature only causes harm. There is something very deep at play here—the sense of self-worth needs to be found within.

Venus Retrograde Statistics

Venus is retrograde for approximately forty out of each 584 days. Theoretically, this means that Venus is retrograde 6.8 percent of the time. Research into databases of about 19,000 charts revealed that only 5 percent of famous individuals listed in them had Venus retrograde. The charts of friends, family, and clients elevated this number to 7 percent.

Out of the 14,161 records with Rodden rating B or above listed in AstroDatabank, 219 have Venus stationary retrograde—Muhammad Ali, George Bush Sr., Rita Coolidge, Zelda Fitzgerald, Nigel Hawthorne, Richard Simmons, and Gianni Versace; and 216 individuals have Venus stationary direct in their charts—Deepak Chopra, Tony Blair, Jean Chrétien, Aldo Gucci, Mata Hari, Cat Stevens, and Julia Parker.

Mars is retrograde 9 percent of the time, Mercury 17 percent, and the outer planets 33-42 percent of the year. Whenever a planet changes sign, direction, or declination, or forms a potent aspect, we astrologers can count on it to reflect change. Likewise, when a person has a planet in a rare condition, there is an increased probability that this person has unique qualities and traits associated with the planet. Variances from the norm result in variances in the normal, expected behavior, attitude, and approach.

Closing Thoughts

Perhaps Venus retrograde individuals prefer the company of animals in a setting close to nature rather than embracing the social, elitist setting in which they were raised. Their senses and sensibilities are highly developed. What truly matters is learning self-love and how to appreciate themselves. In addition, these individuals have a unique appreciation of beauty—seeing it where others do not.

---

1. Tim Sanders, *The Likeability Factor* (New York: Crown, 2005).
2. *Guild News* is the newsletter of the Fraser Valley Astrological Guild.

CHAPTER TEN

Venus in Love

"When we fall in love it can seem a lot like mental illness—a blend of mania, dementia and obsession that could almost be mistaken for psychosis. Now neuroscientists have produced brain-scan images of this fevered activity. In an analysis of the images appearing in the *Journal of Neurophysiology*, researchers in New York and New Jersey argue that romantic love is a biological urge distinct from sexual arousal. The researchers assert that it is closer in its neural profile to drives such as hunger, thirst or drug craving than to emotional states like excitement or affection."[1]

Venus has two distinct sides—romantic love and sexual passion. When we fall under her charms, we forget the natural instincts that normally are engaged in social situations. In the chart these are represented by the seventh house, its occupants, and its ruler. Venus enters a house in our chart and forms aspects to planets in our chart (they could be the seventh-house ones). Her presence in that area instantly stirs an urge within us. Note that when we fall in love, the entire chart is abuzz with frenzied activity. We are focusing primarily on the role Venus plays.

So, assuming that the stage is set, we are under her spell. Let's consider that Venus made her dramatic entrance right through the front door (Ascendant). Pure adrenaline is rushing through our bodies—she is sending bold signals that she desires you and you alone. The

Figure 27—Rose

first house is the natural house of Mars, so the aspects from your Mars or from hers, (often both) work similarly, and voilà—we have chemistry! Venus is pheromones and Mars is adrenaline—the act of falling in love belongs to Mars, and the allure to enchantress Venus. Each house has its unique role and stirs up something deep within us—the fifth is people we instinctively let into our hearts, and the eighth house gives a fated feeling—kismet.

In a man's chart, Venus defines how he can be captivated. Venus is attracted to the essence—not to the feminine and masculine pronouns. As the current Venus "decade" was launched, same-sex unions became a topic of debate—love knows no bounds.

When Venus connects with the Sun, we have an instant bond. Connections are aspects, and the major ones—conjunctions, sextiles, squares, trines, and oppositions—are present between the two charts when we find love. Falling in love is truly illogical and does not mean we are destined to marry. Love and marriage are two very different "animals." We crave that feeling that puts all our senses in a tizzy.

Venus shows what we crave, and Mars what we are willing to do about it. *We most definitely need a strong Mars link both ways*; for example, her Mars trine his Venus, and his Venus sextile her Mars. It is nice to have a harmonious aspect between the two Venuses—it adds mutual rapport, shared values, and ease in accommodating. When we fall under that spell, we don't really know what we've got, nor do we care—the euphoria masks any warning signals. In synastry (comparing two charts for cross aspects), we often hear that we are not compatible with the signs on either side of our Sun, yet marriages between people born thirty to sixty days apart are the most common.[2] In these instances, we often find that there is a strong Sun-Venus connection.

Conjunctions between two planets in synastry—his Venus and her Moon, for example—give an intimate bond. Sextiles between the same give an easy camaraderie—we can talk about it. Squares are a friction aspect, which is feisty, sexual chemistry—agree to disagree. With trines, we share the same mode of operation. Oppositions make us drawn to each other despite the differences—that is the natural connection between Venus and Mars. Oppositions create chemistry but can also make it hard to be together when the passion is spent. We love but may not like, and need to be willing to kiss and make up after the battle of wills. Her Venus opposite his Venus tends to be the hardest aspect to accept—their values clash! We can find workarounds with harsh aspects between Venus and Mars—the chemistry behind it is exhilarating.

Note that Venus by progression (the maturation process) travels through three to five signs in your chart. So by age twenty-five, at the latest, your Venus will have moved into the next sign.[3] Through estimation, Venus moves through a sign via progression at intervals of twenty-four to twenty-five years. We desire different things as we age. Our progressed Venus is very active when we fall in love. Mars and the luminaries tend to be the most active at the time we encounter the object of our adoration. Venus in each sign desires something different.

Venus in Aries

Love is not something you feel. It's something you do.
—David Wilkerson

Venus in Aries is hunting Mars. She is a rebellious, nonconforming go-getter; someone strong, quick to act, a bad boy or a tomboy. She knows Mars's likes and dislikes; what makes him tick—so he is easy prey. She is a physical, enthusiastic spark plug, who instantly ignites flames of passion. All you can do is her bidding—wild, spirited, exciting action with no time to think. She wants to be shown that she is loved. She loves surprises and only seems to want to be shown that you appreciate all those impetuous gestures. If Venus in Aries tires of Mars, she is gone in a literal flash—she isn't a possession.

A woman with Venus in Aries wants to initiate the relationship. Wooing, sentimental cards or insipid tokens of affection are not enough. She is not clingy and does not need to be taken care of; she needs both a friend and a lover. She craves demonstrations of affection, because she is rather romantic. Do it, don't just talk about it!

A Venus in Aries man loves his sports and the thrill of speed. If it has an engine, he's in love already. He isn't looking for a clingy, high-maintenance gal. He dislikes anything humdrum. He'll encourage you try anything once; *he* certainly will. He might not instantly know what you love and like, but he is a quick study—you only have to show him once. He'll treat you as his equal; it wouldn't occur to him that you are not.

The challenge here is that you can't slow down or slouch. Venus in Aries needs to always go, go, go!

Venus in Taurus
Don't settle for the one you can live with; wait for the one you can't live without.
—Anonymous

Venus in Taurus yearns for someone to grow old with in comfort—she can be sensible in love. She loves the courting, craves security, and freely depends on her partner to be her source of support. He built me a home—what is a girl to do? Her voice and her touch make you weak in the knees. She's the conjugal goddess of love—cherishing, tender, earthy, and romantic wrapped up in a sensual body. Venus in Taurus is giving—pleased only when you are pleased.

Venus in Taurus likes the courting, dinners, flowers, and gifts, and hates to do without. She is possessive and reluctant to surrender that which is hers. She needs to be valued and appreciated. Taurus is the sign of foreplay. That song from years ago that goes, "I want a lover with a slow hand; I want a lover with an easy touch," must have been about Venus in Taurus. She speaks in a soft, gentle, caressing voice and seldom takes no for an answer. She won't be rushed; she is loyal, stable, and reliable and needs to have all her senses fed.

Venus in Taurus men know how to treat a woman—shower her with gifts and please her. He may have an old-fashioned attitude about a woman's place, but he takes his role as the provider seriously. You can count on his affection, his labors of love. He will protect you and is as solid as a rock.

With a strong Venus connection, we share the same tastes, values, passions, and priorities. We are eager to make love—what a comfortable connection.

The challenge here is to provide stability and security. When those practical needs are met, Taurus is very accommodating and accepting.

Venus in Gemini

Love is like a butterfly, as soft and gentle as a sigh. The multicolored moods of love are like its satin wings. Love makes your heart feel strange inside. It flutters like soft wings in flight. Love is like a butterfly, a rare and gentle thing.
—Dolly Parton[4]

This versatile, inquisitive goddess craves shapeshifting Mercury—a lover and a pal. Her words and pet names for you are like music to your ears. Venus in Gemini is a beguiling, youthful, and intelligent companion. Each day to her is a new life; she'll play different roles and keep life so interesting. She'll flirt with you to show she's still interested. She'll giggle at your jokes and listen intently to your deeper ponderings.

Venus in Gemini is genuinely interested and interesting. Keeping things lively is important to her, and sharing is vital. She prefers to have an extraordinary life. To hold her interest, you need to be fascinating, clever, and the range in between—a friend today, a teacher tomorrow, and so on. She flirts with others because it is fun; jealousy is a turn-off. The way to her heart is through her mind. Once a friend, always a friend, even if the romance fades away. Without mental stimulation, she may escape into her own mind to create a different reality. Like Peter Pan, she doesn't want to age and enjoys her flights of fancy.

Unlike other men, Venus in Gemini men chat, gossip, and know both trivia and facts. Variety is the spice of life, so you are allowed multiple personalities. You can have things in common and do things individually—freshness keeps it lively.

Venus and Mercury sing about love, sending tingling sensations through our bodies. How is it possible that those words in old love songs suddenly sound so beautiful and intimately personal? In love, the simplest words are music to our ears.

The challenge here is obvious: without rapport, easy repartee, and sharing, Venus in Gemini will seek sweeter nectars. If life gets too routine and humdrum, she will flit away.

Venus in Cancer

The best and most beautiful things in the world cannot be seen or even touched.
They must be felt within the heart.
—Helen Keller [5]

The empathetic, demure, caring, and sensitive Venus in Cancer wants the Moon. She is so enticing, with her soft face and those big, watery eyes. She charms by being so sweet, so loving, and so shy. You just want to protect her and ensure she'll always be safe. She's romantic—most of those love quotes were penned by Venus in Cancer. She is looking for someone with whom to nest and build a home. She can mother you or behave like a child. She bonds for life; there is no escaping her "claws."

Venus in Cancer knows that love is completely illogical. She needs to feel that love and learn to trust it. Capturing her heart's attention may take a bit of time. Her affections grow stronger with each passing year. She needs to be romanced, and she'll keep every memento—they are tokens of your affection, and she needs to collect quite a few. Water Venus is ultimately feminine, satin, and lace, designed to catch her knight in shining armor.

A Venus in Cancer man is enchanted by a feminine woman, the kind you can introduce to your mother. He is emotional, chivalrous, caring, and protective—he understands your moods. He'll spend his last dollar on a rose for you. He yearns for a woman to marry and raise a family with. Traditions and family values are important to him.

Love is familial—moonlit nights cast such a romantic spell. Our instinctive responses bring about tears of joy, and the emptiness of being alone is suddenly replaced by feeling complete. We are restless without our love; we cannot eat, drink, or sleep.

The challenge here is that it takes time to bond and learn to trust; jealousy needs to be kept away. The littlest things can hurt. It isn't right until *it feels right*.

Venus in Leo

Joy does not simply happen to us.
We have to choose joy and keep choosing it every day.
—Henri Nouwen

Venus in Leo is glamorous and warm, the one everyone gathers around. She'll entertain and charm. She's lovable and loads of fun. The golden goddess of drama makes a grand entrance; she can't tolerate a cold shoulder. It's champagne and caviar, galas and balls—she is romance à la Hollywood. You'll fetch and carry and bow to her desires, because she gets those flames of passion roaring. She's a born performer at everything she does—a little praise and she'll fetch you the Sun. You deserve the best.

She needs to be the only one—the one adored and cherished. She is alluring, generous, full of praise, and rather high-maintenance. She'll dress to the nines, cherish her body, and decorate the temple she lives in. She deserves the best, and if she falls for you, you'll be made to feel like a king—after all, she is looking for the Sun! She'll prop up your ego and boost your confidence.

The creative Venus in Leo man adores women who know how to look great. He loves kids, sports, and toys, and makes a wonderful father. This is not an easy Venus to catch. He knows how to both play and praise—he, too, is reaching for the Sun. The Sun we know warms, burns, and scorches.

Sun-Venus connections are prominent in love. It's a fervent, burning desire—we feel cold alone. We adore and melt in the presence of this energy. We want the world to see that we and only we are truly in love.

The challenge here might be that because this is such a creative Venus, marriage and ever-lasting love might take a back seat to the passion to create something bigger than life.

Venus in Virgo

They will treasure a diamond in the rough, uncovering the beauty of the multi-faceted diamond within. It is not that something needs to better, but that it can be, which titillates the Virgo.

This intelligent and discerning Venus appraises her choices in her quest for analytical Mercury. She craves someone who is her intellectual equal. She charms with her wisdom, her unceasing interest in you. You may not notice you are being interviewed, because her unwavering attention is focused on you, holding you captive. She is not looking for perfection but potential—if you are too perfect, she loses interest. She craves someone to whom she can teach the wonders of love and life. She doesn't want to hear what you heard, but what you think of it.

Venus in Virgo can readily fall in love with a calling or opt for a working relationship. This is an analytical placement. She likes simplicity, and whatever she ingests needs to be pure—including love. Feelings and the object of her desire are scrutinized, and at times perfection wins out over perfecting. A frequent query of Venus in Virgo is: Do I really love or is it a figment of my imagination? Virgo in Venus can opt to take a lover for pragmatic reasons.

Men with Venus in Virgo may be waiting for Miss Perfect or someone rather close. They are charming, witty, and bright but somewhat timid—not an easy catch.

When we have that Mercury-Venus connection, we have a sense of familiarity coupled with the senses gone wild. Suddenly logic is elusive, and we quiver and stutter. Short-circuitry sends unique signals coursing through our veins.

Once the partner is chosen, there are no obstacles or challenges that cannot be overcome. Venus in Virgo is a devoted, loyal, and accommodating lover, once you have been found to fit the bill. Lest we forget, this is an earth sign, so Venus in Virgo is sensual and rather learned in the ways of pleasuring a partner.

Venus in Libra

True love is eternal, infinite, and always like itself. It is equal and pure, without violent demonstrations: it is seen with white hairs and is always young in the heart.
—Honoré de Balzac[6]

To the ethereal, divine, perfect goddess of charm, love is sacred. Venus in Libra will shower you with gifts—tokens of her affection. She is the classic, elegant lady with perfectly coiffed hair, who turns heads as she suddenly appears. She knows she is beautiful, and everyone else does too. She graces you with her undivided attention, making everyone else dissolve. The moment she gazed at you, you knew your heart was hers. Venus in Libra is so enticing that she merely has to be present to get attention. Her fleeting caress feels like a promise of more—here she is the mighty Aphrodite.

She loves sentimental cards and poems and is delighted with any gift, because it means you were thinking of her. She may not know exactly what you would like, but she will find out. She loves to chat about everything, though not what she paid for something. She is very accommodating and can play different roles for you. She is the one who will walk to you wearing sexy lingerie and let you chase her to the perfectly decorated boudoir, so let her have her silk, satin, and lace. She'll play golf with you if you'll take her to the theater.

At times, she is just as happy to go out with her many friends and let you off the hook. She will wither away without praise and words of love from you.

You fell in love with her charms; she is the one who lets you have your male moments and appears to defer those big decisions to you. You may know that she planted those ideas in your mind, yet you still play along. Men with Venus in Libra know the rules of courtship. They are brave enough to go after Aphrodite herself. They want to adore and be proud of their love; to them, she is without flaw.

With the seventh-house connection or Venus to Venus, we feel like Aphrodite or Adonis—perfect, lovable, and desirable. Our hearts beat faster, and taking a breath or wasting time on anything mundane leaves our minds. There is only love, and nothing else matters.

Living up to those expectations can be challenging. However, Venus in Libra can be most accommodating as long as everything is nice, harmonious, and balanced.

Venus in Scorpio

It's the soul's duty to be loyal to its own desires.
It must abandon itself to its master passion.
—Rebecca West [7]

When intense, magnetic Venus in Scorpio walks into a room, the men turn to look and the women hold onto their men a little tighter. She is looking for her soul mate, and her eyes see right through you. She is stalking Mars and Pluto; she bonds throughout eternity. She demands passion and commitment—with mind, body, and soul. She is intriguing, seemingly holding on to secrets. She allures and mesmerizes with her sex appeal; she is equipped with more pheromones than most. She also knows men intimately; there is nothing or nowhere you can hide.

She may keep her secrets, but you won't get to hold on to yours. She is tempestuous and possessive, and her emotions range from fiery love to icy-cold hate. She needs intensity and potency, and if you are her prey, resistance is futile. She is a powerful ally and an even more powerful enemy. She is compassionate, loyal, and extremely generous with those she deems worthy. She demands your support to pursue her creative passions. Your chivalrous actions will catch her attention.

A man with Venus in Scorpio is scouting for a strong woman, someone who understands his passions, desires, and devotion. He needs someone charismatic and strong, who can offer emotional support. He has powerful feelings and emotions. He won't unleash those until he can trust you.

Venus-Mars chemistry in the night sky or through the eighth house is pure kismet and fate. We gaze into those eyes and are swallowed completely by a sumptuous passion. The world fades away; that burning passion needs to be satiated.

The biggest challenges here are trust and a noncommittal attitude. Love me or hate me; indifference is intolerable. It's impossible to contain the raging ocean, and even a still one is deep. If you are not devoted to the depths of those emotional currents or cannot be trusted, you'll be eliminated.

Venus in Sagittarius

The joy of a spirit is the measure of its power.
—Ninon de Lenclos

Venus in Sagittarius is the seeker searching for Jupiter. Jupiter is all about fun, philosophy, and adventure. It's a big world out there. Venus in Sagittarius may travel the world, looking for love. She is full of life and laughs and likes to goof around. If she can teach you, or you can introduce new and bigger concepts to her, you'll capture her attention for a while. Let the games begin! She is hard to pin down and is born to teach, preach, and enlighten. She can be a heartbreaker; she is the ultimate free spirit, with a yearning for complete freedom to do as she pleases.

She is captivating and somehow larger than life, but capturing her attention requires that you are wise or think that she is. This is a kind of teacher/guru-student relationship. She'll expand your view of the world and introduce you to her interests, from playing sports and loving animals to climbing Mount Everest. Life is a grand adventure, and you are welcome to come along for the ride. Those Venus-in-Sagittarius hugs melt resistance and simply make you fall in love.

The Venus in Sagittarius man is attracted to a woman who stands out in the crowd, an adventurous spirit with a passion for life. If he is sporty, he might want a philosopher. In myth, Jupiter was the benevolent philanderer. However, you are free to stay and free to go.

There is exhilaration, joy, and bliss in the Venus-Jupiter and ninth-house connections. Suddenly, you feel you are more than you thought you could ever be. Someone grand has professed their love, and your feet don't touch the ground.

It's all a challenge with Venus in Sagittarius, but most people who are smitten by the Sagittarian love for life stay to bask in that eternal flame.

Venus in Capricorn
Ambition is like love, impatient both of delays and rivals.
—Sir John Denham

Venus in Capricorn is traditional and appreciates societal values. She is ardent and sensual—but not in public, mind you. With age and practice, she only gets better. She is image-conscious and values her privacy. She is often your classic beauty, with legs that go on forever. Even in terms of looks, she seems to grow younger and more gorgeous with each passing year. Her ageless quality captures our attention. She does not see it herself—to her it is the make-up, the outfit, the perfect light, or your impaired vision. She works, supports, encourages, and makes you feel that you can conquer the world. She may capture the attention of an older man, who from there on out believes he has gone to heaven. She is also pragmatic and materialistic—there are things a gal needs to have.

Venus in Capricorn is mature beyond her years; time, after all, is on her side. She is ambitious and will aid your climb to success. She believes in your ideas, talents, and skills. She talks with conviction—no flattery, but merely honesty and support. You need to have drive and motivation to catch her attention; she doesn't go for flightiness. She is building a relationship, and the best ones take work and effort.

Men with Venus in Capricorn desire a classic beauty, someone traditional, preferably a wife to stand by her man, who can play the part of the "corporate wife." Capricorn, like Virgo, knows how to perfect and to provide. He also oozes sex appeal—it is an earthy Venus. While we associate Saturn with delays, the earthy senses are tough to put on hold.

With a Venus-Saturn or tenth-house connection, it is those earthy senses that drown out any practical considerations. The body is tingling, the knees provide little support, and there is an unquenchable thirst.

The challenge here is that if you cannot be counted on and relied upon to see to your duties, the pragmatic Capricorn Venus may lose some of that faith in you.

Venus in Aquarius

There are two sorts of curiosity—the momentary and the permanent.
The momentary is concerned with the odd appearance on the surface of things.
The permanent is attracted by the amazing and consecutive life
that flows on beneath the surface of things.
—Robert Lynd

Venus in Aquarius is full of startling contrasts. She is looking for someone unique and reliable. She wants an original, someone brilliant and different. She loves her freedom even more than Sagittarius. She can talk the talk, but she won't walk the walk until someone extremely special catches her fancy. She can cultivate a friend into a lover; find one in the chat room or at a distance. After all, she is looking for both Saturn and Uranus neatly wired together.

She will make you feel brilliant in a friendly, utterly charming, and welcoming manner. In love, she can accommodate unique arrangements. She loves groups, social causes, and making others intrigued about the secrets of the universe, human nature, and anything that causes her to wonder. She falls in love very young or rather late—often yearning for the one that got away. She is candid and truthful, so you always know where you stand. Perhaps this is the goddess who in stories took a lover when she wanted one—to be relegated afterward to a special friend.

A man with Venus in Aquarius wants a unique, somewhat unconventional woman with a brilliant intellect, who can understand his thoughts—perhaps simply someone who can understand his brilliance and encourage his yearning to gain more knowledge. Underneath it all, he may still want something conventional. The way to his heart is through his mind.

Uranus is frequently involved when love strikes. When we are in a Uranus shock, we do not seem to know what happened. We cannot understand why we are feeling this way. It might feel like we are plugged into electricity; we are suddenly nervous, excited, and no longer in control of our intellectual faculties. Sometimes it takes a strong force to make us take note.

With Aquarius, the biggest challenge in romance is the timing. Venus in Aquarius often mourns that childhood sweetheart. When we fall in love in later years, our progressed Venus plays a huge role. Venus in Aquarius can bridge unique obstacles.

Venus in Pisces

Dreams come true; without that possibility,
nature would not incite us to have them.
—John Updike [8]

The ethereal Pisces is the exaltation of Venus. It is about both love and spirituality, but also escapism and fantasy. Venus in Pisces yearns for a soul mate in the elusive Neptune—the god of the sea whence we came and where we shall return; the collective unconscious. She is also rather fond of Jupiter, the preacher, teacher, savior, and guru; she has a thirst for wisdom—to give and receive. She grows up early, experiencing passionate love during her teenage or even preteen years. She seems to attract those who need saving. While Venus in Virgo likes to fix and perfect, Venus in Pisces likes to rescue and provide unwavering support. The list of famous people with Venus in Pisces reads like a who's who in the quest for spirituality or in addictions—some are victims and some are saviors.

Venus in Pisces is beautiful, soft, sensitive, empathetic, caring, and kind—the empath. She is ultimately feminine (the one Hollywood loves to stereotype) and the ideal woman—sensitive, emotional, and needing a man to be her support. People with their Sun in Capricorn, Aquarius, Pisces, Aries, or Taurus may have Venus in Pisces. This Venus can be on a quest for something otherworldly; she is the ultimate romantic. She can capture our attention without making an effort. She can also imagine her lover to be what he is not.

A man with Venus in Pisces can be in love with the image of the mermaid. He is the most sensitive and caring; he almost cares too much. He knows women intimately, but coping with any brush-off is difficult. Regardless of what he is like, most women find him sweet—maybe someone to "rescue."

Neptune is the planet of illusion, dreams, and fantasies. When Neptune and Venus connect, we have pure bliss at work. There is a kaleidoscope of colors whirling around; your misty eyes are seeing nothing else but this mystical creature, who stirs all the waters within you. (We are 95 percent water, remember.)

The challenge here might be figuring out how to live up to that ideal and stay on that pedestal.

Closing Thoughts

In the natal chart, if we look at the seventh house, the planets in it, and its ruler, we find a basic description of the spouse. When we look at the fifth house, the planets in it, and its ruler, we have some idea about what we desire in romance. Falling in love and getting married are not the same—marriage takes cooperation and commitment.

Looking at the Venus sign alone does not give us a full description. Astrology is about blending the signs, houses, and aspects. We also mature, so we need to look at those progressed planets. No wonder all of us study this subject continuously, as there is always something new to learn.

Sustaining that "in love" feeling in marriage requires adjustment and cooperation, which Venus handles with exquisite flair—she will get her desires met.

---

1. *Weekend Post*, June 4, 2005; notation from the *New York Times*.
2. French astrologer Didier Castille has done considerable statistical work on marriage and birth dates.
3. This does not apply to you if you were born around the Venus retrograde period.
4. Capricorn Venus conjunct the Sun in the fifth house.
5. Venus in Cancer conjunct the Sun.
6. Venus conjunct Mars in Cancer.
7. Venus in Scorpio.
8. Venus in Taurus.

CHAPTER ELEVEN

Venus and Marriage

Marriage is not a ritual or an end. It is a long, intricate, intimate dance together and nothing matters more than your own sense of balance and your choice of partner.
—Amy Bloom

We do not all marry because we are in love. Arranged marriages in Western culture were rather commonplace less than a century ago. In Finland, up until only a few decades ago, once you turned twenty-four years of age, it was determined that you were a spinster or a bachelor. At that time, the government assigned an extra tax burden to people who were unmarried. Back in the early 1900s that often meant you would be found a husband or a wife. Remember that the right for women to vote and to be considered a person was first implemented in New Zealand[1] in 1893, in Australia[2] in 1902, and in Finland[3] in 1906. In the USA[4] that took place in 1920. The Venus occultation era in Sagittarius at the end of the 1800s began the suffragist movement. In her article "Finland: A Pioneer in Women's Rights," Irma Sulkunen writes: "Seen from an international perspective, the history of women's suffrage contains an interesting paradox: the more vehement the battle, the more meager the results. Countries with the most militant suffragettism had to wait for years, even decades, before they could enjoy the fruits of their struggle, while many small, peripheral countries gave women full parliamentary representation at an early date without much ado."5

Venus has other desires that may override the need to marry for love. She didn't have the perfect husband in mythology either—she was married off. That wedding ring represents eternity but also capture. Marriage today still is legalized by the state or sanctified by the church. Marriage by choice is a fairly new concept. Even when we marry purely for love, the instant we are a couple, we own and share something together.

Because Venus desires more than love, it helps immensely if our Venuses have a nice, harmonious connect—it creates rapport. Venus has a role that we tend to forget: she accommodates and adjusts to get her needs met. In the natural wheel, Venus rules two signs, Taurus and Libra, and the second and seventh houses. The natural aspect between these pairs is a quincunx, and the keyword for this aspect is "adjustment"—some call it an itch you can't scratch. Interestingly, we don't usually realize we have to bend our ways and make room for others until we find ourselves in a situation where we must. In this chapter, we will look at how Venus copes in marriage, where is she willing to compromise, and where she might insist on her way or the highway.

Let's look at those other Venus concepts in quick review. These are simmered down to one word: money. Money buys us possessions, food, and financial security. Venus in the houses, signs, and aspect defines what is the most important. Even when the two nicely placed Venuses are in marriage, those needs vary. We know from experience that financial woes, and differing financial priorities, are big culprits in marital woes.

Further, before we marry, we typically have a proposal of marriage, followed by an engagement. Proposals are a seventh-house issue; we need to come to an agreement. Engagement is intended to be an eighth-house matter—can we commit to the marriage? The actual marriage is a contract, which falls in the tenth house.

Marriage Indicators

Sometimes, marriage is just cute, cuddly, and easy to see. Your seventh-house ruler is in the tenth and you met him at work. His Sun is in your third and communication is a breeze. Your Sun falls in his sixth house and your role is to serve. His Venus is conjunct your Juno[6] (Juno's story is in chapter 13) and you are marriage material. His Venus or Mars is conjunct your Part of Marriage and he instantly wants to marry you. The same patterns are activated in your chart, and you desire the same.

When we meet others, the first thing they see is the Ascendant we have spent years cultivating. It's our façade but also the behavior that has proven to win us acceptance. Others, however, are represented by the opposite house behind our backs, often with qualities we haven't owned. If we have Mars ruling our Ascendant (Aries, Scorpio), then the Descendant is ruled by Venus. We project assertive and even aggressive attitudes and attract a kind and loving soul, one who will put up with our outbursts and strive to create harmony. Conversely, if we have a Venus-ruled Ascendant (Taurus, Libra), our partners are Mars-driven. They incite a passion in us, help us lose inhibitions, and accommodate our need for pleasant, peaceful surroundings, and readily do the hunting and gathering for us.

Mercury-ruled Ascendants have Jupiter ruling the seventh-house cusp (Sagittarius/Pisces). Our horizons expand with these partners; we are likely to travel the world or explore the world of wisdom with our partner. They will encourage us to relax and have a laugh, offering us optimistic support to balance our natural inclinations. If we consider Neptune the ruler of the seventh, we are looking at someone ethereal, romantic, and perhaps idealized as the partner. This can mask our flaws, as Neptune has a penchant for looking at us through those rose-colored glasses, being more focused on seeing the beauty within. Could the Mercury or Venus in a Mercurial sign cope with someone who might like to escape reality, perhaps with the aid of alcohol? Naturally, in marriage, we explain these away in social situations: Well, he drinks because his work is so stressful. She needs to unwind, it was a difficult day for her. Venus is about social pleasantries.

If we have one of the luminaries (Moon/Cancer, Sun/Leo) on our Ascendant, our partners are Saturnine (Capricorn/Aquarius)—reliable, conservative, and perhaps more mature than we are. Our Venus can also make this sound wonderful; we get to see the other side, which we personally appreciate and value. No wonder we talk about our shadow. We tend to get what we don't really expect. Opposites do attract—but they also repel. What brought us together can pull us apart, when the passion fades and we do not know how to adapt and become accustomed to their quirks. The Descendant of the Leo Ascendant is ruled, in modern terms, by Uranus. Well, you like them unique, but you might get eccentric. Does your Venus approve?

The Part of Marriage

Let's consider what factors are incorporated into this Arabic part. The formula for the Part of Marriage is the Ascendant plus the Descendant minus Venus. *Me and You less Venus!* It's quite revealing—we cannot be selfish in marriage nor only want our own desires fulfilled. This ancient calculation removes Venus from the equation.

The Part of Marriage requires a little calculation unless you have a computer program that generates the Arabic parts. The easiest way to accomplish this task is to convert those degrees to the ones they occupy in the 360° wheel; that is, 0° is the beginning of the zodiac in Aries and 360° is the final degree of Pisces. For example, an Ascendant at 14° Leo is the 134th degree of the 360° wheel. It helps to think how many signs of 30° each precede the sign the planet or point is in.

Here is an example:

Ascendant at 5°23' Leo	125°23'
Descendant 5°23' Aquarius	305°23'
Add the two together:	430°46'
Deduct Venus at 16°23' Libra:	196°23'
Part of Marriage:	234°23'
Less full signs 30 x 7 = 210	210°00'
24°23' Scorpio	24°23'

So the Part of Marriage is at 24°23' Scorpio.

That is the Part of Marriage for Prince Charles of England. (His chart was discussed in chapter 3.) Princess Diana's Venus was at 24°24' Taurus. On the day of his second marriage, his progressed Moon was opposing that point at 20° Taurus. Camilla Parker-Bowles's Sun at 24°47' Cancer trines Charles's Part of Marriage. However, his Mars is conjunct her Juno, and he wanted to marry her for decades. On the day of the marriage, the transiting Sun is often in close aspect with this part. Perhaps the Part or Marriage in a man's chart is conjunct the transiting Mars. Venus might sit on your seventh-house cusp.

When we opt for a divorce, we may find that this calculated point is receiving several harsh aspects.

Tim Stephens, a syndicated columnist and astrologer in British Columbia, Canada, casts a marriage part chart by placing the degree of the Part of Marriage as the Ascendant degree, using equal houses. He then places the natal planets in the wheel to predict the kind of partner and the time of marriage. This technique works amazingly well. There is another interesting part worth considering in terms of love and appreciation. The formula is the Ascendant plus Venus minus the Sun—persona enhanced with appreciation or lovability less the ego.

Married Venus

Success in marriage does not come merely through finding the right mate,
but through being the right mate.
—Barnett Brickner

In the first year of marriage, we completely overlook any shortcomings in the other. Those falling-in-love passions override any other needs. We are on our best behavior, using all of the innate charms of our Venus. We might also think we will change the other in order to get all our needs and desires met. Let's think of her Venus in Taurus in the fifth house sextile Jupiter in Pisces in the third. She wants to play, sense that romance, and go to restaurants for meals between sensuous lovemaking sessions. His Venus in Capricorn in the second house, while still charmed and spellbound, begins to worry about the money, how the bills are going to be paid. He begins to suggest that they eat at home more, and start saving for the home and for holidays. Well, she can adjust. Perhaps this escalates, and now she is buying and cooking the food, and he might note that they cannot afford filet mignon every night and asks for simpler meals. She might decide that he is too stingy and make commentary on that. Venus wants what she wants, but she will gradually adjust.

You are in love and married—but can you live together? When our Venus connects with the Ascendant or Descendant of another, we tend to like them. Venus in a nice aspect to the Sun in a potential partner's chart shows that we like who they are. A nice connection to the Moon shows that we understand their emotional nature and can accommodate their emotional needs.

An important aside, the primary connection we need to share our daily lives with another is a harmonious aspect between the two Moons. The Moon represents our habits, routines, and familial settings. That lunar connection helps identify how well our habits and daily routines meld together. Those instinctive responses when our emotional comfort zone is breached are difficult to handle if we do not understand them in our partner.

We can overcome most challenging connections between the two charts—including a difficult connection between the two Venuses; however, in lasting marriage it is best to have a nice connection between the Moons and Venuses. Astrologer Julie Kelly,[7] from New Mexico, points out that Saturn on any of our personal planets is a killer—the constant criticism or "parenting" is something we cannot tolerate. Venus can play that game too, if you do not appreciate the same things she does.

What Might Venus Have in Her Arsenal?

What might be the issue that Venus has a hard time accepting? Venus in Aries might feel trapped if she is not allowed to remain independent to pursue her other interests. Venus might become upset if "we no longer do things together." Venus conjunct Mars might state, "Well, you used to love my spontaneity." Venus by house seems to show how we adjust the easiest in order to accommodate the other—she can pour her passions into that area of life. We can usually also accept the negative qualities of our Venus in another. For example, Venus in Scorpio can adjust to jealousy and possessiveness, and Venus in Gemini can find someone else with whom to exchange thoughts and ideas—without leaving the marriage.

Bringing Out the Worst in Venus

Venus in Aries knows how to bully us and demand more action. Venus in Taurus may show us the flaws in our ways of handling money or food or can chide us about not being productive. Venus in Gemini is quick with both wit and disapproval; if you said you were going to do it, it is instantly in black and white. Venus in Cancer can play the parent or revert to the child; she can hide in the corner and cry over any perceived hurt. A Leonine Venus can be an absolute drama queen, with theatrics and preplanned tears. Venus in Virgo can be critical and find those flaws and tell you to fix them or else. A Libran Venus can hurl insults disguised as praise; it might take days to figure out that you were put in your place. She can twist those words with the best of them. Venus in Scorpio may quickly turn into the Queen of Ice, and let you writhe in your misery for days and leave you to try to figure out what happened. Venus in Sagittarius knows how to preach and blurt out hurtful words. The arsenal of Venus in Capricorn can be rather potent; she remembers facts and figures, and at the opportune moment will give you a running total and can make you feel like an incompetent fool. Venus in Aquarius uses the weapon of shock; lightning hits when you least expect it, and she quickly manages to deliver her truths while making sure you understand that you hurt her feelings. Surely Venus in Pisces is always nice? We all know about the "woe is me" syndrome, but she also holds secrets from way back when; she remembers hurts and knows how to inflict them. She can make you feel guilty by simply making you sense it.

Every marriage goes through stages and phases, and along the way we learn to accommodate new facets of our Venus and their Venus. Venus can be rather uncompromising if the quality, characteristic, or possession she fell in love with and chose to marry in the first place leaves the equation. Let's assume she married to have financial security. If that isn't there for her after all, she may find a job or she may opt to look for a new partner elsewhere. Perhaps she married for camaraderie, and if that wanes, she might settle for having friends to fill that void. If both camaraderie and passion have departed, she is likely to love and leave you.

Closing Thoughts

When we meet that special someone and fall in love, some of the most consistent astrological themes in a chart are a Moon, Venus, or Sun connection by progression and to the natal chart. Often we see the progressed Moon conjunct the natal Sun, Venus, or Mars and/or a personal planet progress to natal Venus, Mercury, or Mars or to the Part of Marriage. The conjunction seems to be the strongest indicator, but oppositions often come into play as well. Venus stationing on the seventh-house cusp and hovering close to it for four months might make us seriously contemplate our marital status. For example, I fell in love with my husband when I had an exact progressed Sun-Moon conjunction (progressed New Moon) on my natal Venus. We married when my progressed Venus was conjunct my North Node. When we marry, Venus likes to be visible on an angle by transit; the Part of Marriage and Juno are also activated by transit.

We are typically married by the time we notice the negative side of Venus—she knows how to play that social role to perfection. The initial passion of courtship and that blissful first year of being newly wed eventually fade. Remember that during the first year the woman remains the bride. Therefore, we need to learn to adjust and accommodate. Venus in Cancer might want domestic bliss, but may marry a Venus in Libra, who also desires lots of social activity. The two can learn to agree and compromise, and live happily ever after—one for you, one for me, and one for us. Alternately, we can bolt and find someone else, who will also bring out the best and worst in us.

In the interim, we get those Venus periods, and some are going to influence our marriage in order to review, reassess, and readjust. Some signs have more staying power than others do—the fixed ones, for example: Taurus, Leo, Scorpio, and Aquarius. I have watched marriages tied with the transiting Sun in one of these signs last and endure, growing more solid with each passing year.

I have been married for a quarter of a century. My Scorpio Venus learned to trust; I no longer experience those moments of intense jealousy when my Leo Sun husband would chat with another woman. I saw flirting, while he was simply being the charming man I fell in love with. His Venus in Cancer in aspect to Pluto and Uranus learned that I would stay. It got stormy at times, but we worked things out. Have we changed? Probably. But more importantly, life has unfolded, and had we not been able to accept changes in each other, we would no longer be together. His businesslike wife became a full-time astrologer on his watch.

I fell in love with Venus as I completed this book; I hope you have more insights into her essence. She is love and appreciation; she is a beneficial energy with the capacity for wrath. Her essence resides in you, within you. She is perfect!

---

1. Independence chart for September 26, 1907, shows Venus in Libra.
2. Based on the Dominion Day, Australia has Venus in Sagittarius conjunct Uranus.
3. The Declaration of Independence chart has Venus in Aquarius.
4. The July 4, 1776, chart shows Venus in Cancer conjunct Jupiter in the seventh using 17:14 as the time in Philadelphia, PA.
5. Irma Sulkunen, "Finland: A Pioneer in Women's Rights," *Virtual Finland*, http://virtual.finland.fi/netcomm/news/showarticle.asp?intNWSAID=25734, June 2000.
6. There are many free software programs available for downloading. Kepler's Starlite calculates the asteroid positions—http://www.astrosoftware.com.
7. http://www.soulpathastrology.com.

CHAPTER TWELVE

Venus and Societal Change

*As man now reasons in dreams, so humanity also reasoned
for many thousands of years when awake.*
—Friedrich Nietzsche

On June 8, 2004, Venus danced across the disk of the Sun. Astronomically, these are called transits. The astrological definition of a celestial body passing in front of another is known as an occultation. No, Venus cannot hide the huge Sun. She is normally hidden from sight during a New Venus. However, over the course of a millennium she performs her "dance" sixteen times—eight times in Gemini and eight times in Sagittarius—always brilliantly visible.

Occultations are similar to eclipses; we know that these take place at the nodes. All of the planets have a slowly regressing nodal axis. These are the intersections of the path of the planet and the ecliptic.[1] Currently, the North Node of Venus is in Gemini and her South Node is in Sagittarius. The most recent rare occultation took place on June 8, 2004, at 17°53' Gemini. Venus works in harmony and in cooperation, so just as every New Venus is paired with a Full Venus, the occultations take place in pairs. The next occultation in Gemini will take place on June 5, 2012, at 15°45' Gemini. The time lapse between each pair of occultations in Gemini is a quarter of a millennium—243 years. Note that the essence of Gemini is to share knowledge and information without adding any analysis to it.

Unlike the nodal axis of the Moon, the nodes of Venus move slowly; therefore, for thousands of years the occultations can only take place in Gemini and Sagittarius. The nodes of Venus move along at approximately 30" per year (similar to the precession of the zodiac), which translates to approximately 8° per millennium. The nodes are currently at 16°44' Gemini/Sagittarius.[2] Thus, they have been in these two zodiac signs since the start of our modern calendar two thousand years ago.

We had two occultations in Sagittarius at the end of the nineteenth century, on December 4, 1874, and December 9, 1882. Venus is about relationships, so let's do it in pairs.

There have been fifty-two transits of Venus across the Sun between 2000 BCE and 1882 CE. The first one recorded by astronomers was in 1639 CE.[3]

Gemini	Sagittarius	
1032 May 24	1145 Nov 25	
1040 May 22	1153 Nov 23	
1275 May 25	1388 Nov 25	
1283 May 23	1396 Nov 23	
1518 May 25	1631 Dec 07	Note that the Sun currently moves into Gemini around
1526 May 23	1639 Dec 04	May 20.
1761 Jun 06	1874 Dec 09	
1769 Jun 03	1882 Dec 06	
2004 Jun 08	2117 Dec 10	
2012 Jun 05	2125 Dec 08	
2247 Jun 11	2360 Dec 12	
2255 June 09	2368 Dec 10	
2490 Jun 12	2603 Dec 15	
2498 Jun 10	2611 Dec 13	
2733 Jun 15	2846 Dec 16	
2741 Jun 13	2854 Dec 14	

This data is from the NASA website. The dates 1153 November 23 and 1396 November 23 were not listed. As with all the other cycles, these repeat in degrees.

Looking the table, we can see that the time difference between the first occultation in the pair and the first one in the successive one is 243 years. The time lapse between the last one in Gemini and the first one in Sagittarius is 105 years. Once again, we encounter the hundred-year sleep.

We all agree that eclipses at the South Node have a different feel and impact than ones at the North Node, when we are looking at the dance between our planet, the Sun, and the Moon. There are fewer variables with the Venus "eclipses," as the nodes do not change; the North Node is ruled by Mercury and the South Node by Jupiter. In a compass, the needle always points to the magnetic north; it is our guiding direction. The axis of the world runs north and south, and we currently pin the shifting position of the North Pole to the North Star, Polaris. The Southern sphere is likened to a foundation, subjectivity, and things of personal nature. The Sun is in the northern declination in the Northern Hemisphere for the summer solstice. This discussion can get complicated, so as a reminder we flip our charts around for the two spheres of the world. We also place the winter solstice sign at the top of the chart in our natural wheel. I see this as our chart being a reflection of the heavens.

Gemini Occultations

The path of Venus is incredibly dynamic. Her role is not merely to define beauty and to please, but she also has her "decades," when she is instrumental in defining societal change.

Timelines: 2004–2012, 1761–68; 1518–26; and 1275–1283.

1760s
- The Industrial Revolution had just begun.
- The occultation in 1761 led to the creation of the AU, astronomical unit. Astronomers could define the distances and the size of our universe by determining the distance between the Sun and Earth with accuracy. Astronomers traveled to various places on the Earth to observe the event. Using Venus was apparently a suggestion of Edmond Halley (1656–1742), who did not experience a Venus transit in his lifetime. Johannes Kepler had suggested using Mercury.

- On May 22, 1761, the first life insurance policy in the United States was issued in Philadelphia.[4]
- We had just experienced a Uranus-Neptune mutual reception, which is once again happening at the beginning of the twenty-first century. The inventions at the beginning of the Industrial Revolution era were brought to common people. Venus is associated with sharing and relationships through her association with Libra and the seventh house in the natural wheel.
- In the first decade of the twenty-first century, the Venus and Uranus-Neptune mutual reception phenomenon repeats from a similar pattern in the 1750s and 1760s. However, we now have an accelerated and condensed period where the two stellar events occur in the same time period—fast progress and social reform.
- *The Social Contract* by Jean-Jacques Rousseau, a book of political philosophy, was published. The concepts in this book are like those in the American Declaration of Independence.
- There were numerous wars fought during these times that centered on land rights. Lands were gifted by nations who no longer found it worthwhile to hold on to them. Canada, for example, was ceded to Britain by France in 1761.
- In the USA, the Mason-Dixon line was put in place.
- Taxation was implemented on the colonialists by Britain, starting with an innocent postage stamp.
- The group Daughters of Liberty was formed. Slavery became an issue, and the last witch hunts were going on in Sweden.

1518–1526

- In 1517, Martin Luther founded the Lutheran Church, and in the course of the Venus decade brought religion to the common people. He also wrote the first ABC book, allowing regular people access to the written word and thus knowledge. He grew to believe that faith alone was humankind's link to the justice of God, and believed that there was no need for the vast infrastructure of the Church.
- Nanak founded Sikhism, a combination of Buddhism and Hinduism. Titian painted *Offering to Venus*.
- In 1526, William Tyndale, priest and translator, completed and published the first complete version of the New Testament in English at Worms, Germany.
- Henry VIII appealed to the Pope for permission to divorce Catherine of Aragon.

Other Notations

- Many of the outspoken individuals who made a huge difference in our global society were born during these Venus occultation years: Somerset Maugham, 1874; Guglielmo Marconi, 1874; Gertrude Stein, 1874; Herbert Hoover, 1874; Quincy Adams, 1767; Andrew Jackson, 1767; Franklin D. Roosevelt, 1882 (natal Mars retrograde in Gemini opposite the occultation); and Harry S. Truman, 1884. First Ladies Dolley Madison and Elizabeth Monroe were born on either side of the occultation in 1768. Samuel Wilson was born in 1766, and while historians are not in complete agreement, the prevailing theory is that Uncle Sam was named after Samuel Wilson.
- Following the 1275–1283 Venus occultations, Scotland made it legal for women to propose to men by passing a leap year act in 1288.
- During each occultation period, a significant work of art dedicated to Venus was crafted.
- In 1033, there was a huge pilgrimage to Jerusalem to mark the 1,000th anniversary of the crucifixion of Christ; 1032–1040 was a Venus occultation "decade."
- In 1526, Magellan completed the first circumnavigation of the world—the world was a globe!
- Catherine the Great came to power in Russia in 1762. Mary, Queen of England (1516), was born in the Occultation Decade.
- At the time, we had the first occultation in Gemini in our lifetime (2004), so much previously hidden knowledge hit the mainstream. Venus spoke through many authors about feminine wisdom. We began to gain insights into her nature as an initiator to wisdom. The astronomical highlights were broadcast. In Canada, many astrologers were featured in the media discussing actual astrology and not merely horoscopes.
- Based on current carbon dating, the structure at Newgrange in Ireland and the Rosslyn Chapel (which was integral in Dan Brown's novel *The Da Vinci Code*) were both built during Venus "decades."

Sagittarius Occultations

Timelines: 1631–1639, 1874–1882; December 2117 and 2125.

- In 1874, a battle between people without jobs and the police in New York left hundreds injured. There was a depression in the USA following the Civil War.
- 1874–1882: postal services, radio, long-distance phone calls, global communication, labor unrest, and riots.
- In 1874, Levi Strauss began marketing blue jeans with copper rivets—copper is the metal associated with Venus, and the jeans have their connection to the working classes. The Sagittarius themes appear to be about the rights of common people.
- In 1884, Greenwich Mean Time and time zones were established.
- In the 1140s, European leaders outlawed the crossbow with the intention to end war for all time.
- In the 1140s, Gratian wrote *Decretum*, a standard treatise on canon law in Bologna.
- In the 1150s, a map of western China was printed; it is the oldest-known map.
- In the 1150s, the first foreign currency exchange contracts and government bonds were issued in Italy.
- In the 1380s, Charles VI declared an end to taxes forever.
- In 1386, the University of Heidelberg was founded.
- The 1390s was a time of expulsion of Jews from France.
- In 1396, the last Christian Crusade ended in disaster.
- On June 20, 1397, the Kalmar Union united Denmark, Sweden, and Norway under one monarch. The Union lasted until 1523 (Venus decade in Gemini, 1518–1526).

Venus in Gemini loves information and talking, but it is deeper than just that. Gemini is about words and Venus about the right words, so imagine how a creative a writer that energy could produce. Gemini is the sign we like to associate with ideas. Numerous significant events define the 1760s, but the gist of it is that people fought to ensure everyone had access to information, and the fight for those rights took place during the Sagittarius periods. Discoveries and books written made the information and concepts accessible to the masses.

We are in for a lot of social change. Our information age becomes the information speedway; even sacred texts from most cultures are available free of charge on the World Wide Web. Money and taxation have historically been prominent issues during these periods. Globally, this could result in the euro becoming the comparative currency for the world, ousting the U.S. greenback. New currencies have appeared during these periods. Religions have been altered significantly—split into two different groups or blending two belief systems into one. People have fought for their rights to equality and equity. Perhaps we have reached a period when science and religious beliefs come a little closer together. In the previous periods, the inventions of the strong Uranus periods became household items with the Venus period. Now, once again, these two celestial events are entwined.

We know there are incredible inventions out there being discovered at lightning speed, and these will become household items in the next Venus years:

- New weapons and weapons technology emerge.
- Science and religion blend together into a new evolutionary concept.
- Trade practices change dramatically.
- Medical breakthroughs occur at an overwhelming pace.
- Science becomes more accessible to the masses.
- Philosophy and occult knowledge are written for the common person.
- Travel and methods of travel change rapidly.
- Hybrid vehicles emerge into the mainstream.
- Technological advances gain practical application quickly.
- New religions emerge and debate over the freedom of faith takes place.
- The world learns to walk the walk of diversity.

We are already seeing signs of change. Spirituality is a thread in best-selling novels. The sweeping sociological changes during the Venus decade will be harder to recognize while they are right before our eyes, as we tend to see history unfold more clearly in hindsight.

We need to remember that these occultations take place while Venus is retrograde. Venus is about our values and priorities at both the personal and collective levels. During the past few decades, most of us, especially those of us who are older and have lived a few decades as adults, are becoming discontent with the state of affairs in the world, the humanmade inequality and injustice. This Venus decade is likely to bring changes in how we collectively treat each other. Throughout time, as humans learn more, they become more involved in their destiny and not as much pawns of fate. Collectively, we are in for lessons in humanity, and about to learn not to hoard what we feel is rightfully only ours.

The main lesson Venus has to teach us is of love, but we tend to confuse this with mortal love and the appreciation of the good things in life. It is easier to neglect the fact that we are all connected as one great big whole, and when one part ails, the whole entity suffers.

In hindsight, this is a wonderful, auspicious period, during which we will gain new appreciation for life and all things living. We are more ready to share what we know to help others reach new levels of awareness. Information is becoming much more accessible to everyone.

What is truly interesting is the fact that Venus remains visible during this transit. Remember that this event gave the astronomical unit of measurement, so that we learned to gauge the size of our universe. Therefore, the blemishes or wrongs to be righted stay visible during this transit. We are talking about Gemini, the sign that can always find a way to relay knowledge and information. Gemini will always find a way to communicate to ensure everyone has the information.

---

1. The ecliptic is the apparent annual path of the Sun through the zodiac.
2. 2005—heliocentric node; the geocentric node is at 25°26', which is the Anti-Galactic Center.
3. NASA website, "Transits of Venus," http://sunearth.gsfc.nasa.gov/eclipse/transit/catalog/VenusCatalog.html.
4. http://www.decades.com is a wonderful source for historical dates and data. Many of the references in this chapter are sourced from this site.

CHAPTER THIRTEEN

Other Feminine Archetypes

In addition to Venus, the Moon is the other major feminine factor in the astrological chart. The Moon represents mothering and nurturing. She is what we look to for our instinctive responses, the mother, home, and family. She represents emotional security and our changing moods. The best way to understand her ways is to watch the Moon wax and wane for a few months. There are enough books on her to warrant skipping her in this chapter. The progressed Moon is a wonderful timer in relationships; when she comes to our Venus, we seem to fall in love. The eclipse cycle lasts eighteen years, during which the apogee point of the Moon travels through the zodiac twice. That point is known as Lilith—the one Venus chased away from under her tree of wisdom. Venus has thirteen stations and conjunctions with the Sun over the course of eight years. The Moon has thirteen New and Full Moons over the course of a year. None of the cycles works in isolation, nor can we always clearly ascertain which of the goddesses belongs only to Venus or only to the Moon. The feminine is mysterious.

The four major asteroids—Ceres, Vesta, Pallas Athena, and Juno—were discovered during the first two years of the nineteenth century. At that time, astronomers considered them planets, and the textbooks of the time reflect this. In light of the fact that these four were major deities on Olympus, maybe it is a shame that they were not granted the status of planets. The cycle of the four major asteroids lasts about five years, and they are situated between Mars and Jupiter. The asteroids provide us with valuable clues, and in terms of mythology, they used to hold higher offices than Venus in the hierarchy of ancient gods and goddesses.

Ceres ⚳

The symbol for the asteroid Ceres is a stylized sickle; a crescent of receptivity resting on a cross of matter. It is an upside-down version of the Saturn glyph. The sickle is a crescent-shaped, sharp tool on a short handle, which was used for cutting grass and crops. Sickle cells are associated with red blood. In Greek mythology, Ceres was known as Demeter. Often a minor Roman deity grows to encompass the qualities of an earlier Greek personification. This asteroid moves retrograde for about three months at intervals of about fifteen to sixteen months.

Ceres is the Earth Goddess and patroness of fertility—not to take anything away from Venus, who is also linked to fertility or more precisely to the pregnancy cycle. In myth, Ceres is the mother of Persephone, who was abducted by Hades to become his bride. Mother Ceres in her grief caused the Earth to grow barren. The eventual resolution was that Persephone would get to spend the crop months with her mother on Earth and would return to spend a third of the year (half a year in some versions) in the underworld with her husband. This is linked to the months when the Earth is dormant and the soil is regenerating.

It is interesting to note that the same abduction-type theme or a descent to the underworld is in most myths. It seems to be the role of the feminine archetypes to venture into the unknown to face their inner demons either to rescue men or their daughters. Ceres has a strong connection with agriculture and farming. Demeter, the daughter of Gaia (Mother Earth) was said to have taught humans how to farm the land. In the Olympian zodiac, she is associated with Virgo, and is often depicted as a woman holding grain, much like the image we now associate with the sign Virgo. She is also one of the twelve main gods and goddesses. Ceres is also associated with nourishment, nutrition, mundane tasks and chores, and with nursing and caring for others.

As we have ceased to be an agricultural society, where food was not processed to the extent that it is today, we can easily see the connection to allergies becoming more prevalent. Ceres is linked to allergies and sensitivities to our environment. In mythological stories,

Ceres is said to have looked after her children without the help of their father. Caring for children has been the primary task of women throughout centuries, and while that is changing somewhat now, it is something to keep in mind when looking at Ceres in the chart. Where do you *nurture* mostly on your own?

Perhaps the role of Ceres is to teach us to live in harmony with Mother Earth, and when we do not, we get to experience ailments directly linked to our environment and food. Many men who have Sun conjunct Ceres are very nurturing and caring, often being the primary parent for their children, and are not averse to cooking, cleaning, and the other menial tasks associated with providing care to another. Moon conjunct Ceres seems to be prominent in the charts of health care workers, especially in alternative health care. As with any single aspect in astrology, when a planet or planetoid is prominent in the chart, the influence is more pronounced.

In terms of career, Ceres is linked to helping and service-type work and professions; it seems to be on the Midheaven for people who have also chosen to call such work their avocation.

Perhaps the questions to ask would be, where do you provide care and who do you tend to look after? Do bear in mind that the symbol for Ceres, the sickle, is used to harvest and to bundle. It was a symbol on the flag of the former Soviet Union. Mother Russia, which was the term used, looked after its people, who did not need to do without the necessities in life. The luxury items and trinkets that we truly do not need in order to survive used to create those long lines on the streets of Russia. So look at Ceres by house to see where you can wield your sickle to harvest the plentiful crops.

Perhaps we should also start taking note of Ceres for fertility. Ceres in strong aspect to natal Venus seems to be a pronounced theme for conception. You need to know people very intimately to have the actual conception chart. In one case, the woman's natal Sun in an earth sign received a partile trine from Ceres, which was conjunct the Moon the night she conceived. Sometimes we astrologers are asked about good conception times, so adding Ceres to our arsenal may be helpful. Look to Ceres also for menopause and eliminating harmful items from your diet; Ceres is practical, giving us what we need but not all that we necessarily desire—that is Venus's role.

Vesta ⚶

The astrological symbol for Vesta looks like a stylized flame in a hearth, perhaps reminding us of the eternal flame in Rome, which was kept burning permanently as a symbol of the belief that Rome was eternal. The flame was attended by six Vestal Virgins—meaning chaste and pure. In Greece, Vesta was known as Hestia. The four major asteroids[1] were discovered when the industrial revolution was in full motion and women were starting to enter the workforce. The asteroids are associated with feminine archetypes, and the role of women in society was rapidly changing. Astrology draws heavily from Greek and Roman mythology, where each of the asteroids has a personified deity associated with it. We have to assume that synchronicity is at play rather than these asteroids being named by chance.

Hestia is associated with diligent work in the affairs of both home and state. She is the goddess of home and hearth, who refused marriage. She was virtuous and highly regarded in both Roman and Greek times as the protector of home and family, perhaps the most beloved of all the deities. She was the first child of Cronos (Saturn) and Rhea[2] and the elder sister of Zeus (Jupiter). Her genealogy makes her one of the original gods and titans, and we have relegated her to the rank of an asteroid; yet the gods assigned to the Sun, Moon, Mars, Jupiter, Uranus, Neptune, and Pluto are a part of her family tree.

Vesta is self-disciplined and controlled, creative, modest, and plain. She seems to be service-oriented. She has a mission, which is to keep something alive. Wherever we have Vesta situated in our chart is where we want to keep that flame going. Naturally, when Vesta is at the angles or in aspect to the angles, she is more prominent and more readily or visibly expressed by us. When Vesta is not in aspect with the balance of the chart, she may lack visible expression. When Vesta is at an angle in a marriage chart, the marriage seems to work well and both partners support each other.

When Vesta is positioned at either the MC or zenith,[3] we have a productive, work-oriented person, who seldom complains about her workload. (We know that rules have exceptions.) These individuals are reluctant to be idle and are always busily working on that task list, never erasing the difficult ones.

Where Vesta is seems to supply clues to the area in your life in which you are willing to be the one to take care of things. Looking after joint finances might be one expression of Vesta in the eighth; in the fourth, looking after the home; in the tenth, at the zenith, or on the MC, perhaps being the main breadwinner, never shirking your obligations. Where

Vesta is situated shows an area of life that naturally seems to be your responsibility. Look at Vesta in your chart to see how it is linked and if you indeed naturally assume the responsibility of keeping the issues of that house working smoothly. Oh, and by the way, you do not get credit or praise for doing those particular chores, nor should you expect to.

Vesta retrogrades once every eighteen months for three months. Incidentally, the cycle length is the same as that of the lunar nodes through the signs and Venus through her Morning and Evening Star phases. Vesta spends approximately three months traveling through a sign. A colleague and friend of mine has Vesta in her third house in Pisces and considers herself the Keeper of Words and Knowledge. With Vesta in the ninth house, we may find ourselves sharing philosophical concepts that have the potential to allow others to broaden their horizons, armed with a different perspective.

The Moon by transit and progression connects with all the points in our chart, providing us with observational knowledge about how these energies operate. My first reminder of the role of Vesta in the astrological chart was loud and clear when my progressed Moon crossed over natal Vesta in Virgo at the zenith of my chart. I actually looked at this in hindsight, as at the time I was just too busy working. The days were not long enough, and in addition to working at my job, I was studying accounting. I was also finding time to give astrological lectures, write a column, and tend to my large family. At no point did I feel anyone was taking advantage of me or that I was a slave to others. I was happy to be doing so much. That was also when it first dawned on me I might be a bit of a workaholic, at my happiest when I was being productive. Learn about your Vesta by looking at what happens when slow-moving planets form a conjunction to your natal Vesta.

Is this same type of information expressed through other planets and points in the chart? Absolutely. Every concept is described numerous times in the horoscope through different planets, signs, houses, and aspects, so that we understand the message.

Juno ⚹

Prior to the Church declaring St. Valentine's Day, after posthumously granting sainthood to one Valentine, the Pagan holiday marking the 14th of February was in honor of Juno (Februata), the queen of the Roman gods and goddesses. In mythology, Juno was the faithful wife of philandering Jupiter and the mother of Mars. The Greeks knew her as Hera. The symbol we use for Juno is a stylized star with seven rays resting on the cross of matter.

When we study relationship issues in the astrological chart, we look primarily to the fifth, seventh, and eighth houses and to Venus and Mars. Venus, over the centuries, has come to represent so many concepts that it becomes rather challenging to analyze them all objectively. In Juno, we have another archetype that can yield wonderful insights into our marriage-type relationships.

Venus represents our desires, what we like and dislike. In a man's chart, Venus represents the ideal woman. When we look at the Moon in a man's chart, we are also looking at the concept of motherhood, and thus how the partner would potentially be in a family setting. With the Moon, we are also considering the habits of the person; thus, if your Moons are not in harmony, your daily lives may be a source of strife. In a woman's chart, we primarily look to Mars for the ideal man, while blending the indications of the Sun into the mix. In both instances, we also need to look closely at Juno. Juno in the eleventh house might indicate that it is important for you to marry someone who is also a friend and confidant. If Juno is in the second house, we might draw the conclusion that the spouse needs to have the same set of values and priorities. Juno in a fixed sign might be a strong indication for mating for life. Juno, unlike Venus, did not have many partners; rather, she was committed to being a wife and making the arrangement work.

Juno is an administrator par excellence; she has what it takes to run an efficient household, business, and enterprise, especially if the partner (like Jupiter in myth) is out there gallivanting. This asteroid seems to be active by transit and progression at the time of marriage as well as when we change residence. Juno was faithful, but also jealous—she would inflict revenge on the women rather than Jupiter. Women who have sons learn to understand men differently. What a weapon Juno has in her arsenal, being the mother of the male—Mars.

Pallas Athena ⚹

The astrological symbol for Pallas is the diagonally placed square atop the cross of matter. Diamonds are often associated with courage and daring. In meteorology, diamond shapes can sometimes signify clear air, and in chemistry, soft soap. It has been used in biological contexts for an individual of unknown or not stated sex. The symbol for Pallas Athena is one of the alchemists' signs for sulfur, which is also a chemical associated with Venus.

Pallas Athena—Minerva, to the Greeks—was born fully grown and clad in glorious armor from the august head of Zeus. According to myth, the wise Titaness Metis, whose name means thought, might have been her mother. In some versions, Zeus had her without the help of a woman; thus, we often hear of her association as a daddy's girl. Pallas is the strategist and skilled instructor, ready to do battle for what is fair and just—with a strong connection to law. The animal associated with Pallas is the owl, which we consider a symbol of wisdom. This bird of prey can turn its head a full 180° and has exceptional vision and hearing. It symbolizes wisdom and the ability to see and hear clearly despite the darkness.

Pallas is also considered the patron of the web, and in myth she taught the art of weaving as well as industry, justice, and skill. She was a wise advisor to her father. A Gorgon-monster (or Medusa) capable of turning her enemies into stone has been inscribed onto her shield. Athena taught people how to use the wheel, auger, and axe and how to make sails and different kinds of instruments. Philosophy, poetry, and speech were other skills with which she was credited. A young, mortal girl named Arachne, who was a talented weaver, challenged Athena to a weaving contest. Arachne lost, and Athena turned her into a spider. This is one indication of the potential connection to competition and even vengeance associated with this asteroid.

Minerva/Pallas won the competition in which the gift that would benefit man the most was being sought. Neptune offered a horse as a gift, and Pallas Athene the olive tree. In her honor, the Greek capital was named Athens. Pallas was the protector of the nation/state of Athens. The other origin of Pallas is dual; according to one story, Pallas was a Titan, a giant, whom Athena killed and thus got the second part of her name. The other version is that Pallas was Athena's playmate who was accidentally killed. In Greek, the word pallas means girl, which could simply be an additional indication of her virginity.

Pallas Athena is considered a warriorlike virgin goddess, a sort of Amazon, and was possibly considered a leader of a matriarchy. While on one hand she was considered masculine, a daddy's girl, on the other she was wisdom personified. She was the only goddess on Olympus who was associated with the feminine intellect and strength of mind. In this regard, she is the symbol of women's liberation. Pallas is also connected with justice, law, and creative intellect in general. The cooperation of mind and body could be the gist of it.

Lilith ⚸

The symbol for Lilith is a black Moon crescent resting on the cross of matter. The European ephemerides routinely include her listing, and more and more astrologers are beginning to use this in the chart. Her mythology is varied. Lilith is likened to the first Eve in biblical terms. She was the one who refused to be submissive to Adam. Some liken her to the temptress in Eden, the snake, which represents wisdom as well as the cycle of life—birth, death, and transformation. She is typically depicted as half-woman and half-bird, with talons for feet. She is said to screech in the night as she flies. Many an amulet has been fashioned to ward off her effect as the taker of children.

She is rebellious, demands equality, and is seen as the archetype of a demon and "bad woman." Lilith sounds like the first independent feminine spirit. She can mark the place in the natal chart where you refuse to submit to the demands of others. She is also symbolic of wisdom; her bird, like Athena's, is an owl.

The first female president of Finland, Tarja Halonen, was born on December 24, 1943, at 19:20 in Helsinki, Finland.[4] She is known as an independent spirit with strong views, who uses colorful language and lives according to her own code of ethics.

My father had Lilith conjunct the Ascendant in the twelfth house in Aries. He raised me as if I were his son. His favorite statement to me (but not to my younger sisters) was: "You can do anything a man can do, indeed better. Well, standing at the toilet to take care of business might be challenging, but you can train." He opened my eyes to the invisible realms through the predictions of Nostradamus and various books on spirituality. I also have Lilith conjunct my Ascendant, but in the first house, and I have always considered myself equal with men. I am not a women's libber—it never occurred to me that I needed to be. Equality is a given.

---

1. Meaning "starlike" and referred to as planetoids. The asteroid belt is between Mars and Jupiter, and the four major asteroids are the largest in size.
2. Rhea is a satellite of Saturn.
3. The zenith is the point directly above us and always at a 90° angle from the Ascendant—very significant in terms of our work rather than avocation.
4. Data comes from AstroDatabank—from memory. Submitted by Finnish Astrologer Markku Manninen via e-mail.

Acknowledgments

Writing this book about Venus has been an absolute blessing. I gained new insights and hope that you did as well. I have met such wonderful people during the process, each willing to share their personal experience with the timing of the Venus cycles and of how Venus works for them.

People used to worship Venus and erect temples for her. We can tap into her essence to create inner harmony. I cannot even count how many blessings she has brought me, from friendship to beautiful Venusian gifts. I salute the perfection within all of us, and want to mention a few last thoughts and special people in my life.

Which connections does your Venus have? I was raised by Venus in Taurus parents and have a Leo and Aquarius Venus sister. I married a Venus in Cancer and raised four children whose Venus placements are Virgo, Sagittarius, Pisces, and Cancer. Venus in Aries is a popular placement with my friends, with a few Venuses in Capricorn, Libra, Leo, and Gemini tossed into the mix. Think through your own connections, and you'll likely find them all. However, when it comes to the Sun signs, your list is likely to be missing quite a few. Venus is simply more accommodating.

Writing takes long hours and incredible commitment—not just from the writer but those supporting the effort. My husband, whose second-house Venus cooks for me as long as I am being productive, made sure I was eating and worried if I would meet those deadlines. Look to the second house for time. Without his love and support, this book would never have seen the light of day. After all these decades, I still adore him.

My best friend and colleague, Janice Brown, has Venus in Aries in the third house conjunct my third-house Moon. She read chapters repeatedly and offered wonderful ideas, new words, and never-ending encouragement. She listened to my ideas for hours on end. Janice's frail health did not stop her from being intimately involved in the birthing of this book. I am so blessed to have such a wonderful friend in my life.

Carolyn Zonailo, a dear friend, poet, and colleague, has a fifth-house Venus in Sagittarius conjunct my twelfth-house North Node. She offered stylistic ideas and thoughts on the concepts, listened, encouraged, and gave me a crash course in writing, including how to use that wonderful em dash. We spent hours going over what needs to be in this book and what should go in another book.

My presubmission copy editor, Maggie Frost, is a delightful Australian (as is my husband) who made sure the typos did not reach the publisher. Maggie's Venus is in Cancer in the first house and retrograde. We spent hours on the phone and exchanged numerous e-mails. She would often ask me what else she could do to make the process easier. I made a new friend.

I bonded instantly with Donna Van Toen, Canadian astrologer and author, at her annual conference, where I gave a Venus lecture. Donna's Venus is in Pisces in her second house, which trines my Sun and falls in my third house. Her Mars is trine my Venus on the Midheaven—a natural to review the book and write a foreword, provided she liked the material. (Yes, my Venus-Saturn worried.) I am simply honored.

My Venus in Virgo son read a couple of early draft chapters, and critiqued them and gave some wonderful ideas, which helped set the tone. My Scorpio son with Venus in Sagittarius has been inspired by the passion that went into writing this book. His Scorpio Sun was captured by an enchanting Venus in Scorpio girlfriend. Venus in Scorpio warms up slowly, but while I wrote about the goddess, the two of us with Venus in Scorpio have truly bonded. My Venus in Pisces son would fetch me things and run errands, so I could keep on working.

My Venus in Cancer daughter, with her Mars on my Sun, sent her friends to see me about their love lives and other issues. She also brought me my puppy, Winnie, with Venus in Aries in the fifth house, when she moved away with her own puppy, who had been my "office" dog for two years.

I also want to mention all my students and clients, who have been so encouraging and supportive. They shared their personal, intimate Venus stories—I am thankful.

In 1991, I was the founding president of the Fraser Valley Astrological Guild and am once again its president. All of the members have offered wonderful support and encouragement, and I want to thank them all. The Guild is a Venus in Virgo conjunct Jupiter in the fourth house—we are family.

Finding the patterns of planets in an ephemeris is not easy. David Cochrane of Cosmic Patterns added the capability to generate data on the cycles of Venus in Kepler 7.0, at my request, which made it a breeze. The names of famous people are primarily from Astro-Databank software. The tools modern astrology has are simply marvelous; both software makers provided unsurpassed technical support.

I also want to mention Stephanie Clement of Llewellyn, who enthusiastically cornered me at the 2003 ISAR conference—I was working on an astrological primer. This lady, with Venus in Capricorn and Mars conjunct my Venus, pried out that I loved the Venus cycles, and before I knew it, my journey to unearth Venus and find the right words to describe her essence had begun.

Who would you want to be your editor when the subject matter is charming Venus in her many guises? Venus has a way of illustrating her energy in these matters as well. Andrea Neff, with her Venus in Libra, made sure that Venus spoke with both beauty and clarity. When I needed something changed again, she would calmly state not a problem—Venus in Libra is accommodating. Venus is Libra makes things look good; any omissions in the book are all mine. Chances are, however, that there aren't any.

Bright Blessings and my gratitude,
Anne Massey
Fort Langley, British Columbia, Canada
July 1, 2005

Appendix I: Astrological Data

- Venus rules Taurus and Libra.
- The Moon is exalted in Taurus; Saturn is exalted in Libra.
- Venus's houses in the natural wheel are the second and seventh.
- Venus is in her Joy in the fifth house.
- Venus is exalted in Pisces.
- Venus is fallen in Virgo. (A planet opposite its sign of exaltation is in its fall.)
- Venus is in her detriment in Aries and Scorpio—Mars's signs. (A planet opposite the sign it rules is in its detriment.)
- Venus moves through a sign in an average of twenty-five days.
- Venus returns to the same point in the zodiac at intervals ranging from 292 days to 410 days.
- The speed of Venus is 1°15' per day close to the Full Venus.
- Venus is conjunct the Sun at nine-month intervals.
- The inferior conjunction—New Venus—takes place when Venus is retrograde.
- Venus is at perigee, or closest to the Earth, when retrograde. (All planets are.)

- Venus is at aphelion, or farthest from the Sun, when retrograde.
- The superior conjunction—Full Venus— takes place when Venus is direct in motion and at apogee (farthest from the Earth) but closest to the Sun (perihelion).
- Venus is never farther than 48° from the Sun.
- Venus has phases just like the Moon.
- Venus spends nine months as a Morning Star.
- Venus spends nine months as an Evening Star.
- Venus completes a full cycle around the Sun in 584 days, or approximately eighteen months.
- Eight Earth years equal 2,920 days, and five Venus years equal 2,920 days.
- Venus is retrograde for approximately forty days at eighteen-month intervals.
- The Venus retrograde period begins twenty days before the inferior conjunction and ends twenty days after it.
- The successive Venus retrograde and direct stations in each of the five zodiac signs—within approximately 2.5° of the previous one—occur at eight-year intervals.
- The successive Venus inferior and superior conjunctions in each of the five zodiac signs—within approximately 2.5° of the previous one—occur at eight-year intervals. Example: In June 2004, Venus stationed retrograde at 26°06' Gemini, and in June 2012, Venus will station retrograde at 24° Gemini.
- Venus is invisible to the naked eye for approximately twelve days around the New Venus conjunction.
- At the inferior conjunction, a thin crescent remains in view through a telescope.
- The rare occultations take place at 121.5-year intervals, at which time Venus crosses the disk of the Sun in full view.
- The pair of occultations take place both at the North and South Nodes of Venus.
- The nodes of Venus move at the same rate as the precession of the zodiac—approximately 32" per year. (Each of the ages is about 2,600 years long.)
- There are four pairs of occultations in Gemini per millennium—June 1518 and 1526, June 1761 and 1768, June 2004 and 2012, and June 2247 and 2256.
- There are four pairs of occultations in Sagittarius per millennium—December 1388 and 1396, December 1631 and 1639, December 1874 and 1882, and December 2117 and 2125.

- The radical return of Venus—the return of Venus to the same point in any point of its cycle in relation to the Sun—takes eight years.
- In classical astrology, the years ruled by Venus number eight.
- At forty-year intervals, Venus returns to the same position as in the natal chart and in the same phasal relationship with the Sun.

Appendix II: Venus Cycles Data

182   APPENDIX II

Aries

Venus SR	°	'	New Venus	°	'	Venus SD	°	'	Full Venus	°	'
Mar 30, 1929	8	Tau 3	Apr 20, 1929	29	Ari 48	May 11, 1929	21	Ari 39	Apr 21, 1933	1	Tau 4
Mar 27, 1937	5	Tau 50	Apr 17, 1937	27	Ari 41	May 9, 1937	19	Ari 26	Apr 19, 1941	28	Ari 51
Mar 25, 1945	3	Tau 35	Apr 15, 1945	25	Ari 41	June 5, 1945	17	Ari 12	Apr 16, 1949	26	Ari 36
Mar 23, 1953	1	Tau 20	Apr 13, 1953	23	Ari 7	May 4, 1953	14	Ari 58	Apr 14, 1957	24	Ari 20
Mar 20, 1961	29	Ari 49	Apr 10, 1961	20	Ari 51	May 2, 1961	12	Ari 44	April 12, 1965	22	Ari 3
Mar 18, 1969	26	Ari 49	Apr 8, 1969	18	Ari 36	Apr 29, 1969	10	Ari 30	Apr 9, 1973	19	Ari 47
Mar 16, 1977	24	Ari 34	Apr 6, 1977	16	Ari 20	Apr 27, 1977	8	Ari 15	Apr 7, 1981	17	Ari 28
Mar 13, 1985	22	Ari 18	Apr 3, 1985	14	Ari 5	Apr 25, 1985	6	Ari 0	Apr 4, 1989	15	Ari 10
Mar 11, 1993	20	Ari 1	Apr 1, 1993	11	Ari 49	Apr 22, 1993	3	Ari 44	Apr 2, 1997	12	Ari 51
Mar 9, 2001	17	Ari 44	Mar 30, 2001	9	Ari 32	Apr 20, 2001	1	Ari 28	Mar 31, 2005	10	Ari 31
Mar 6, 2009	15	Ari 28	Mar 27, 2009	7	Ari 16	Apr 17, 2009	29	Pis 12	Mar 28, 2013	8	Ari 11
Mar 4, 2017	13	Ari 9	Mar 25, 2017	4	Ari 57	Apr 15, 2017	26	Pis 55	Mar 26, 2021	5	Ari 51
Mar 2, 2025	10	Ari 50	Mar 23, 2025	2	Ari 39	Apr 13, 2025	24	Pis 38	Mar 23, 2029	3	Ari 29
Feb 27, 2033	8	Ari 32	Mar 20, 2033	0	Ari 22	Apr 10, 2033	22	Pis 21	Mar 21, 2037	1	Ari 6
Feb 25, 2041	6	Ari 12	Mar 18, 2041	28	Pis 3	Apr 8, 2041	20	Pis 3	Mar 18, 2045	28	Pis 44
Feb 22, 2049	3	Ari 53	Mar 15, 2049	25	Pis 44	Apr 5, 2049	17	Pis 45	Mar 16, 2053	26	Pis 19
Feb 20, 2057	1	Ari 33	Mar 13, 2057	23	Pis 24	Apr 3, 2057	15	Pis 26	Mar 11, 2061	23	Pis 54

All Times GMT

Gemini

Venus SR	°	'	New Venus	°		'	Venus SD	°		'	Full Venus	°		'
May 29, 1964	6 Can	52	June 19, 1964	28	Gem	38	July 11, 1964	20	Gem	21	June 20, 1968	29	Gem	8
May 27, 1972	4 Can	45	June 17, 1972	26	Gem	30	July 9, 1972	18	Gem	13	June 18, 1976	27	Gem	4
May 24, 1980	2 Can	35	June 15, 1980	24	Gem	20	July 6, 1980	16	Gem	3	June 15, 1984	24	Gem	58
May 22, 1988	0 Can	27	June 13, 1988	22	Gem	12	July 4, 1988	14	Gem	55	June 13, 1992	22	Gem	53
May 20, 1996	28 Gem	18	June 10, 1996	20	Gem	3	July 2, 1996	11	Gem	46	June 11, 2000	20	Gem	48
May 17, 2004	26 Gem	9	June 8, 2004	17	Gem	53	June 29, 2004	9	Gem	38	June 9, 2008	18	Gem	43
May 15, 2012	23 Gem	59	June 6, 2012	15	Gem	45	June 27, 2012	7	Gem	29	June 6, 2016	16	Gem	36
May 13, 2020	21 Gem	50	June 3, 2020	13	Gem	36	June 25, 2020	5	Gem	20	June 4, 2024	14	Gem	30
May 10, 2028	19 Gem	42	June 1, 2028	11	Gem	26	June 22, 2028	3	Gem	11	June 2, 2032	12	Gem	24
May 8, 2036	17 Gem	32	May 30, 2036	9	Gem	16	June 20, 2036	1	Gem	1	May 31, 2040	10	Gem	16
May 6, 2044	15 Gem	22	May 27, 2044	7	Gem	7	June 18, 2044	28	Tau	51	May 28, 2048	8	Gem	9
May 4, 2052	13 Gem	13	May 25, 2052	4	Gem	57	June 15, 2052	26	Tau	42	May 26, 2056	6	Gem	2
May 1, 2060	11 Gem	2	May 23, 2060	2	Gem	46	June 13, 2060	24	Tau	32	May 24, 2064	3	Gem	54
Apr 29, 2068	8 Gem	52	May 20, 2068	0	Gem	37	June 11, 2068	22	Tau	23	May 21, 2072	1	Gem	45
Apr 27, 2076	6 Gem	41	May 18, 2076	28	Tau	26	June 8, 2076	20	Tau	13	May 19, 2080	29	Tau	37
Apr 24, 2084	4 Gem	31	May 16, 2084	26	Tau	16	June 6, 2084	18	Tau	3	May 17, 2088	27	Tau	28
Apr 22, 2092	2 Gem	20	May 13, 2092	24	Tau	5	June 4, 2092	15	Tau	51	May 15, 2096	25	Tau	17

All Times GMT

Leo (Virgo)

Venus SR	°	′	New Venus	°		′	Venus SD	°		′	Full Venus	°		′
Aug 1, 1991	7	Vir	19	Aug 22, 1991	29	Leo	15	Sept 13, 1991	21	Leo	1	Aug 23, 1987	29	Leo	36
July 30, 1999	5	Vir	8	Aug 20, 1999	27	Leo	2	Nov 11, 1999	18	Leo	47	Aug 21, 1995	27	Leo	30
July 27, 2007	2	Vir	57	Aug 18, 2007	24	Leo	51	Sept 8, 2007	16	Leo	36	Aug 18, 2003	25	Leo	23
July 25, 2015	0	Vir	46	Aug 15, 2015	22	Leo	39	Sept 6, 2015	14	Leo	23	Aug 16, 2011	23	Leo	18
July 23, 2023	28	Leo	36	Aug 13, 2023	20	Leo	28	Sept 4, 2023	12	Leo	13	Aug 14, 2019	21	Leo	11
July 20, 2031	26	Leo	26	Aug 11, 2031	18	Leo	17	Sept 1, 2031	10	Leo	1	Aug 12, 2027	19	Leo	7
July 18, 2039	24	Leo	15	Aug 8, 2039	16	Leo	7	Aug 30, 2039	7	Leo	50	Aug 9, 2035	17	Leo	2
July 15, 2047	22	Leo	6	Aug 6, 2047	13	Leo	57	Aug 28, 2047	5	Leo	41	Aug 7, 2043	14	Leo	56
July 13, 2055	19	Leo	56	Aug 4, 2055	11	Leo	47	Aug 25, 2055	3	Leo	30	Aug 5, 2051	12	Leo	51
July 11, 2063	17	Leo	47	Aug 1, 2063	9	Leo	37	Aug 23, 2063	1	Leo	20	Aug 3, 2059	10	Leo	47
July 8, 2071	15	Leo	38	July 30, 2071	7	Leo	28	Aug 21, 2071	29	Can	10	July 31, 2067	8	Leo	42
July 6, 2079	13	Leo	28	July 28, 2079	5	Leo	17	Aug 18, 2079	27	Can	0	July 29, 2075	6	Leo	37
July 4, 2087	11	Leo	20	July 25, 2087	3	Leo	8	Aug 16, 2087	24	Can	51	July 27, 2083	4	Leo	34
July 2, 2095	9	Leo	11	July 23, 2095	0	Leo	59	Aug 14, 2095	22	Can	42	July 25, 2091	2	Leo	30

All Times GMT

Scorpio (Libra)

Venus SR	°	'	New Venus	°	'	Venus SD	°	'	Full Venus	°	'
Nov 9, 1906	14 Sag	45	Nov 30, 1906	7 Sag	6	Dec 20, 1906	29 Sco	27	Nov 26, 1910	3 Sag	26
Nov 7, 1914	12 Sag	17	Nov 27, 1914	4 Sag	39	Dec 18, 1914	26 Sco	59	Nov 24, 1918	0 Sag	58
Nov 4, 1922	9 Sag	50	Nov 25, 1922	2 Sag	12	Dec 15, 1922	24 Sco	30	Nov 21, 1926	28 Sco	27
Nov 2, 1930	7 Sag	23	Nov 22, 1930	29 Sco	44	Dec 13, 1930	22 Sco	2	Nov 19, 1934	26 Sco	0
Oct 30, 1938	4 Sag	57	Nov 20, 1938	27 Sco	16	Dec 10, 1938	19 Sco	34	Nov 16, 1942	23 Sco	30
Oct 28, 1946	2 Sag	30	Nov 17, 1946	24 Sco	50	Dec 8, 1946	17 Sco	7	Nov 14, 1950	21 Sco	3
Oct 25, 1954	0 Sag	4	Nov 15, 1954	22 Sco	24	Dec 5, 1954	14 Sco	40	Nov 11, 1958	18 Sco	36
Oct 13, 1962	27 Sco	38	Nov 12, 1962	19 Sco	58	Dec 3, 1962	12 Sco	13	Nov 9, 1966	16 Sco	10
Oct 20, 1970	25 Sco	13	Nov 10, 1970	17 Sco	33	Dec 1, 1970	9 Sco	47	Nov 6, 1974	13 Sco	44
Oct 18, 1978	22 Sco	48	Nov 7, 1978	15 Sco	7	Nov 28, 1978	7 Sco	20	Nov 4, 1982	11 Sco	19
Oct 15, 1986	20 Sco	24	Nov 5, 1986	12 Sco	42	Nov 26, 1986	4 Sco	53	Nov 1, 1990	8 Sco	57
Oct 13, 1994	18 Sco	1	Nov 2, 1994	10 Sco	18	Nov 23, 1994	2 Sco	25	Oct 30, 1998	6 Sco	32
Oct 10, 2002	15 Sco	36	Oct 31, 2002	7 Sco	53	Nov 21, 2002	0 Sco	3	Oct 27, 2006	4 Sco	11
Oct 8, 2010	13 Sco	14	Oct 29, 2010	5 Sco	30	Nov 18, 2010	27 Lib	40	Oct 25, 2014	1 Sco	49
Oct 5, 2018	10 Sco	50	Oct 26, 2018	3 Sco	7	Nov 16, 2018	25 Lib	15	Oct 22, 2022	29 Lib	27
Oct 3, 2026	8 Sco	30	Oct 24, 2026	0 Sco	45	Nov 14, 2026	22 Lib	52	Oct 20, 2030	27 Lib	7
Sept 30, 2034	6 Sco	8	Octo 21, 2034	28 Lib	22	Nov 11, 2034	20 Lib	28	Oct 18, 2038	24 Lib	47
Sept 28, 2042	3 Sco	47	Oct 19, 2042	26 Lib	1	Nov 9, 2042	18 Lib	5	Oct 15, 2046	22 Lib	28
Sept 25, 2050	1 Sco	27	Oct 16, 2050	23 Lib	40	Nov 6, 2050	15 Lib	42	Oct 13, 2054	20 Lib	10

All Times GMT

Capricorn (Aquarius)

Venus SR	°		'	New Venus	°		'	Venus SD	°		'	Full Venus	°		'
Jan 3, 1974	11	Aqu	22	Jan 23, 1974	3	Aqu	30	Feb 13, 1974	25	Cap	48	Jan 22, 1978	1	Aqu	52
Dec 31, 1981	8	Aqu	54	Jan 21, 1982	1	Aqu	3	Feb 10, 1982	23	Cap	22	Jan 19, 1986	29	Cap	18
Dec 29, 1989	6	Aqu	26	Jan 18, 1990	28	Cap	35	Febr 8, 1990	20	Cap	55	Jan 17, 1994	26	Cap	44
Dec 26, 1997	3	Aqu	56	Jan 16, 1998	26	Cap	7	Feb 5, 1998	18	Cap	28	Jan 14, 2002	24	Cap	7
Dec 24, 2005	1	Aqu	28	Jan 13, 2006	23	Cap	40	Feb 3, 2006	16	Cap	1	Jan 11, 2010	21	Cap	32
Dec 21, 2013	28	Cap	59	Jan 11, 2014	21	Cap	12	Jan 31, 2014	13	Cap	34	Jan 9, 2018	18	Cap	57
Dec 19, 2021	26	Cap	30	Jan 9, 2022	18	Cap	43	Jan 29, 2022	11	Cap	5	Jan 6, 2026	16	Cap	22
Dec 16, 2029	24	Cap	2	Jan 6, 2030	16	Cap	16	Jan 26, 2030	8	Cap	38	Jan 4, 2034	13	Cap	46
Dec 14, 2037	21	Cap	31	Jan 4, 2038	13	Cap	46	Jan 24, 2038	6	Cap	9	Jan 1, 2042	11	Cap	13
Dec 12, 2045	19	Cap	2	Jan 1, 2046	11	Cap	17	Jan 21, 2046	3	Cap	40	Dec 29, 2049	8	Cap	37
Dec 9, 2053	16	Cap	32	Dec 30, 2053	8	Cap	49	Jan 19, 2054	1	Cap	13	Dec 27, 2057	6	Cap	1
Dec 7, 2061	14	Cap	3	Dec 27, 2061	6	Cap	20	Jan 16, 2062	28	Sag	44	Dec 24, 2065	3	Cap	28
Dec 4, 2069	11	Cap	33	Dec 25, 2069	3	Cap	51	Jan 14, 2070	26	Sag	15	Dec 22, 2073	0	Cap	52
Dec 2, 2077	9	Cap	4	Dec 22, 2077	1	Cap	23	Jan 11, 2078	23	Sag	46	Dec 19, 2081	28	Sag	18
Nov 29, 2085	6	Cap	36	Dec 20, 2085	28	Sag	55	Jan 9, 2086	21	Sag	17	Dec 16, 2089	25	Sag	45
Nov 27, 2093	4	Cap	6	Dec 17, 2093	26	Sag	25	Jan 6, 2094	18	Sag	47				

All Times GMT

Appendix III:
New and Full Venus Listings

Date	GMT	°	Sign	'	
July 8, 1900	11:05	15	Can	47	New
May 1, 1901	2:02	10	Tau	0	Full
February 14, 1902	22:57	25	Aqu	20	New
November 29, 1902	2:36	5	Sag	57	Full
September 17, 1903	21:12	23	Vir	47	New
July 8, 1904	8:13	15	Can	42	Full
April 27, 1905	9:50	6	Tau	27	New
February 14, 1906	9:38	24	Aqu	48	Full
November 30, 1906	5:18	7	Sag	6	New
September 15, 1907	1:36	21	Vir	4	Full
July 6, 1908	3:30	13	Can	38	New
April 28, 1909	17:50	7	Tau	47	Full
February 12, 1910	12:22	22	Aqu	56	New
November 26, 1910	13:53	3	Sag	26	Full
September 15, 1911	11:56	21	Vir	31	New
July 6, 1912	2:31	13	Can	38	Full
April 25, 1913	1:48	4	Tau	15	New
February 11, 1914	20:36	22	Aqu	18	Full
November 27, 1914	17:35	4	Sag	39	New
September 12, 1915	18:19	18	Vir	53	Full
July 3, 1916	19:56	11	Can	30	New
April 26, 1917	9:27	5	Tau	34	Full
February 10, 1918	1:45	20	Aqu	32	New
November 24, 1918	1:05	0	Sag	57	Full
September 13, 1919	2:49	19	Vir	16	New
July 3, 1920	20:51	11	Can	34	Full
April 22, 1921	17:37	2	Tau	2	New
February 9, 1922	7:16	19	Aqu	47	Full
November 25, 1922	5:57	2	Sag	12	New
September 10, 1923	11:01	16	Vir	42	Full
July 1, 1924	12:20	9	Can	21	New
April 24, 1925	1:11	3	Tau	20	Full
February 7, 1926	15:08	18	Aqu	6	New

Date	GMT	°	Sign	'	
November 21, 1926	12:27	28	Sco	26	Full
September 10, 1927	17:51	17	Vir	1	New
July 1, 1928	15:31	9	Can	31	Full
April 20, 1929	9:25	29	Ari	48	New
February 6, 1930	17:39	17	Aqu	14	Full
November 22, 1930	18:16	29	Sco	44	New
September 8, 1931	4:11	14	Vir	33	Full
June 29, 1932	4:39	7	Can	12	New
April 21, 1933	16:20	1	Tau	6	Full
February 5, 1934	4:23	15	Aqu	42	New
November 19, 1934	0:19	25	Sco	59	Full
September 8, 1935	8:48	14	Vir	46	New
June 29, 1936	9:43	7	Can	26	Full
April 18, 1937	1:13	27	Ari	36	New
February 4, 1938	4:04	14	Aqu	42	Full
November 20, 1938	6:30	27	Sco	16	New
June 26, 1940	21:13	5	Can	4	New
April 19, 1941	7:34	28	Ari	51	Full
February 2, 1942	17:32	13	Aqu	15	New
November 16, 1942	12:09	23	Sco	30	Full
September 6, 1943	0:05	12	Vir	32	New
June 27, 1944	3:57	5	Can	22	Full
April 15, 1945	16:44	25	Ari	20	New
February 1, 1946	14:19	12	Aqu	8	Full
November 17, 1946	19:01	24	Sco	50	New
September 3, 1947	14:23	10	Vir	14	Full
June 24, 1948	13:37	2	Can	55	New
April 16, 1949	22:48	26	Ari	36	Full
January 31, 1950	6:40	10	Aqu	50	New
November 14, 1950	0:01	21	Sco	3	Full
September 3, 1951	15:08	10	Vir	18	New
June 24, 1952	22:17	3	Can	18	Full
April 13, 1953	8:15	23	Ari	7	New

Date	GMT	°	Sign	'	
January 30, 1954	0:17	9	Aqu	34	Full
November 15, 1954	7:26	22	Sco	24	New
September 1, 1955	7:57	8	Vir	7	Full
June 22, 1956	6:09	0	Can	47	New
April 14, 1957	13:39	24	Ari	20	Full
January 28, 1958	19:47	8	Aqu	24	New
November 11, 1958	12:20	18	Sco	36	Full
September 1, 1959	6:23	8	Vir	5	New
June 22, 1960	16:24	1	Can	13	Full
April 10, 1961	23:51	20	Ari	51	New
January 27, 1962	10:19	7	Aqu	0	Full
November 12, 1962	20:06	19	Sco	58	New
August 30, 1963	1:30	5	Vir	59	Full
June 19, 1964	22:40	28	Gem	38	New
April 12, 1965	4:21	22	Ari	3	Full
January 26, 1966	8:37	5	Aqu	57	New
November 9, 1966	0:40	16	Sco	10	Full
August 29, 1967	21:40	5	Vir	51	New
June 20, 1968	10:22	29	Gem	8	Full
April 8, 1969	15:10	18	Ari	36	New
January 24, 1970	20:27	4	Aqu	27	Full
November 10, 1970	8:49	17	Sco	33	New
August 27, 1971	18:54	3	Vir	51	Full
June 17, 1972	15:09	26	Gem	30	New
April 9, 1973	19:13	19	Ari	47	Full
January 23, 1974	21:19	3	Aqu	30	New
November 6, 1974	13:09	13	Sco	44	Full
August 27, 1975	13:11	3	Vir	39	New
June 18, 1976	4:36	27	Gem	4	Full
April 6, 1977	6:29	16	Ari	20	New
January 22, 1978	6:15	1	Aqu	52	Full
November 7, 1978	21:34	15	Sco	7	New
August 25, 1979	12:38	1	Vir	44	Full
June 15, 1980	7:27	24	Gem	20	New

Date	GMT	°	Sign	'	
April 7, 1981	9:22	17	Ari	28	Full
January 21, 1982	10:06	1	Aqu	3	New
November 4, 1982	2:02	11	Sco	19	Full
August 25, 1983	4:35	1	Vir	26	New
June 15, 1984	22:32	24	Gem	58	Full
April 3, 1985	22:00	14	Ari	5	New
January 19, 1986	16:05	29	Cap	18	Full
November 5, 1986	10:16	12	Sco	42	New
August 23, 1987	6:25	29	Leo	36	Full
June 13, 1988	0:00	22	Gem	12	New
April 4, 1989	23:29	15	Ari	10	Full
January 18, 1990	22:42	28	Cap	35	New
November 1, 1990	15:15	8	Sco	57	Full
August 22, 1991	20:21	29	Leo	15	New
June 13, 1992	16:30	22	Gem	53	Full
April 1, 1993	13:11	11	Ari	49	New
January 17, 1994	2:04	26	Cap	44	Full
November 2, 1994	23:12	10	Sco	18	New
August 21, 1995	0:04	27	Leo	30	Full
June 10, 1996	16:19	20	Gem	3	New
April 2, 1997	13:45	12	Ari	51	Full
January 16, 1998	11:18	26	Cap	7	New
October 30, 1998	4:22	6	Sco	32	Full
August 20, 1999	11:58	27	Leo	2	New
June 11, 2000	10:31	20	Gem	48	Full
March 30, 2001	4:16	9	Ari	32	New
January 14, 2002	11:32	24	Cap	7	Full
October 31, 2002	12:06	7	Sco	53	New
August 18, 2003	18:04	25	Leo	23	Full
June 8, 2004	8:43	17	Gem	53	New
March 31, 2005	3:30	10	Ari	31	Full
January 13, 2006	23:59	23	Cap	40	New
October 27, 2006	17:50	4	Sco	11	Full

APPENDIX III

Date	GMT	°	Sign	'	
August 18, 2007	3:41	24	Leo	51	New
June 9, 2008	4:18	18	Gem	43	Full
March 27, 2009	19:24	7	Ari	16	New
January 11, 2010	21:06	21	Cap	32	Full
October 29, 2010	1:10	5	Sco	30	New
August 16, 2011	12:08	23	Leo	18	Full
June 6, 2012	1:09	15	Gem	45	New
March 28, 2013	17:05	8	Ari	11	Full
January 11, 2014	12:24	21	Cap	12	New
October 25, 2014	7:31	1	Sco	49	Full
August 15, 2015	19:21	22	Leo	39	New
June 6, 2016	21:49	16	Gem	36	Full
March 25, 2017	10:17	4	Ari	57	New
January 9, 2018	7:01	18	Cap	57	Full
October 26, 2018	14:16	3	Sco	7	New
August 14, 2019	6:07	21	Leo	11	Full
June 3, 2020	17:43	13	Gem	36	New
March 26, 2021	6:57	5	Ari	51	Full
January 9, 2022	0:47	18	Cap	43	New
October 22, 2022	21:17	29	Lib	27	Full
August 13, 2023	11:15	20	Leo	28	New
June 4, 2024	15:33	14	Gem	30	Full
March 23, 2025	1:07	2	Ari	39	New
January 6, 2026	16:36	16	Cap	22	Full
October 24, 2026	3:44	0	Sco	45	New
August 12, 2027	0:20	19	Leo	7	Full
June 1, 2028	10:00	11	Gem	26	New
March 23, 2029	20:11	3	Ari	29	Full
January 6, 2030	13:17	16	Cap	16	New
October 20, 2030	11:12	27	Lib	7	Full
August 11, 2031	3:00	18	Leo	17	New
June 2, 2032	9:07	12	Gem	24	Full
March 20, 2033	16:05	0	Ari	22	New

APPENDIX III

Date	GMT	°	Sign	'	
January 4, 2034	2:10	13	Cap	46	Full
October 21, 2034	17:03	28	Lib	22	New
August 9, 2035	18:40	17	Leo	2	Full
May 30, 2036	2:24	9	Gem	16	New
March 21, 2037	9:15	1	Ari	6	Full
January 4, 2038	1:26	13	Cap	46	New
October 18, 2038	1:40	24	Lib	47	Full
August 8, 2039	19:02	16	Leo	7	New
May 31, 2040	2:24	10	Gem	16	Full
March 18, 2041	6:46	28	Pis	3	New
January 1, 2042	12:16	11	Cap	13	Full
October 19, 2042	6:29	26	Lib	1	New
August 7, 2043	12:39	14	Leo	56	Full
May 27, 2044	18:42	7	Gem	7	New
March 18, 2045	22:23	28	Pis	44	Full
January 1, 2046	13:35	11	Cap	17	New
October 15, 2046	16:01	22	Lib	28	Full
August 6, 2047	11:03	13	Leo	57	New
May 28, 2048	19:51	8	Gem	9	Full
March 15, 2049	21:29	25	Pis	44	New
December 29, 2049	21:52	8	Cap	37	Full

Appendix IV: Four Pentagrams

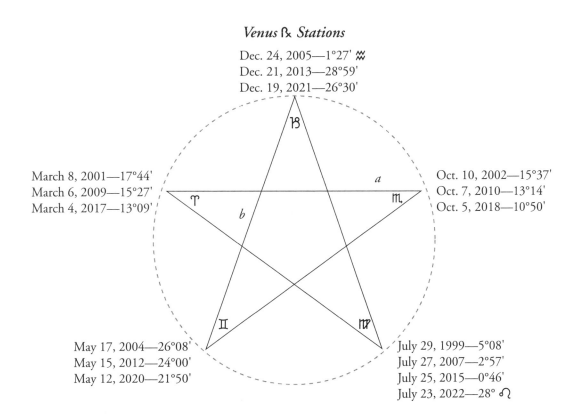

Venus ℞ Stations

Dec. 24, 2005—1°27' ♒
Dec. 21, 2013—28°59'
Dec. 19, 2021—26°30'

March 8, 2001—17°44'
March 6, 2009—15°27'
March 4, 2017—13°09'

Oct. 10, 2002—15°37'
Oct. 7, 2010—13°14'
Oct. 5, 2018—10°50'

May 17, 2004—26°08'
May 15, 2012—24°00'
May 12, 2020—21°50'

July 29, 1999—5°08'
July 27, 2007—2°57'
July 25, 2015—0°46'
July 23, 2022—28° ♌

Appendix IV

Venus Direct Stations

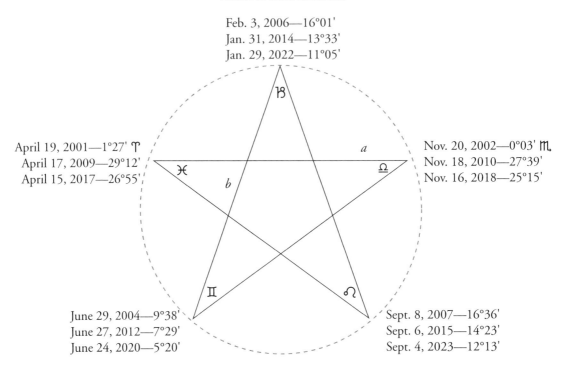

APPENDIX IV 197

New Venus—Venus ℞ ☌ Sun

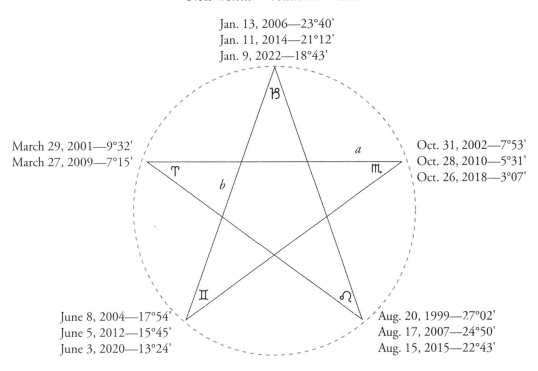

Jan. 13, 2006—23°40'
Jan. 11, 2014—21°12'
Jan. 9, 2022—18°43'

March 29, 2001—9°32'
March 27, 2009—7°15'

Oct. 31, 2002—7°53'
Oct. 28, 2010—5°31'
Oct. 26, 2018—3°07'

June 8, 2004—17°54'
June 5, 2012—15°45'
June 3, 2020—13°24'

Aug. 20, 1999—27°02'
Aug. 17, 2007—24°50'
Aug. 15, 2015—22°43'

APPENDIX IV

Full Venus—Venus ☌ Sun

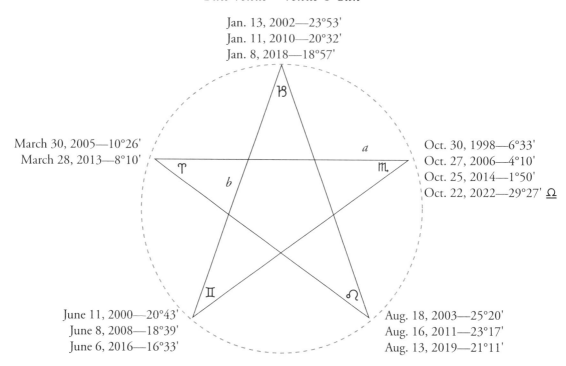

Jan. 13, 2002—23°53'
Jan. 11, 2010—20°32'
Jan. 8, 2018—18°57'

March 30, 2005—10°26'
March 28, 2013—8°10'

Oct. 30, 1998—6°33'
Oct. 27, 2006—4°10'
Oct. 25, 2014—1°50'
Oct. 22, 2022—29°27' ♎

June 11, 2000—20°43'
June 8, 2008—18°39'
June 6, 2016—16°33'

Aug. 18, 2003—25°20'
Aug. 16, 2011—23°17'
Aug. 13, 2019—21°11'

Bibliography

Brown, Dan. *The Da Vinci Code.* New York: Doubleday, 2003.

Dreyer, Ronnie Gale. *Venus: The Evolution of the Goddess and Her Planet.* Aquarian, 1994.

Grant, Michael, and John Hazel. *Who's Who in Classical Mythology.* 1973. Reprint, New York: Routledge, 2003.

Haich, Elisabeth. *The Initiation.* Santa Fe, NM: Aurora Press, 2000.

Kaldera, Raven. *MythAstrology.* St. Paul, MN: Llewellyn Publications, 2004.

Knight, Christopher, and Robert Lomas. *Uriel's Machine: The Ancient Origins of Science.* London: Arrow Books, 2000.

———. *The Hiram Key.* London: Arrow Books, 1996.

McCann, Maurice. *The Sun & the Aspects.* London: Tara Astrological Publications, 2002.

Rudhyar, Dane. *The Lunation Cycle: A Key to the Understanding of Personality.* Aurora Press, 1986.

Sanders, Tim. *The Likeability Factor.* New York: Crown, 2005.

Schaup, Susanne. *Sophia: Aspects of the Divine Feminine Past & Present.* York Beach, ME: Nicolas-Hays, 1997.

Wolkstein, Diane, and Samuel Noah Kramer. *Inanna, Queen of Heaven and Earth: Her Stories and Hymns from Sumer.* New York: Harper & Row, 1983.

Zonailo, Carolyn. *The Goddess in the Garden.* Victoria, B.C.: Ekstasis Editions, 2002.

To Write to the Author

If you wish to contact the author or would like more information about this book, please write to the author in care of Llewellyn Worldwide and we will forward your request. Both the author and publisher appreciate hearing from you and learning of your enjoyment of this book and how it has helped you. Llewellyn Worldwide cannot guarantee that every letter written to the author can be answered, but all will be forwarded. Please write to:

Anne Massey
℅ Llewellyn Worldwide
2143 Wooddale Drive, Dept. 0-7387-0991-3
Woodbury, Minnesota 55125-2989, U.S.A.

Please enclose a self-addressed stamped envelope for reply, or $1.00 to cover costs. If outside U.S.A., enclose international postal reply coupon.

Many of Llewellyn's authors have websites with additional information and resources. For more information, please visit our website at
http://www.llewellyn.com